MATHTOPIA PRESS | GRADES 6–12

Competition-aligned Math With Curriculum Support

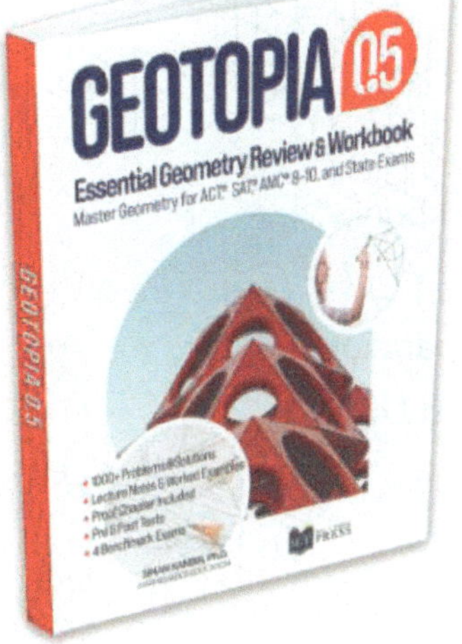

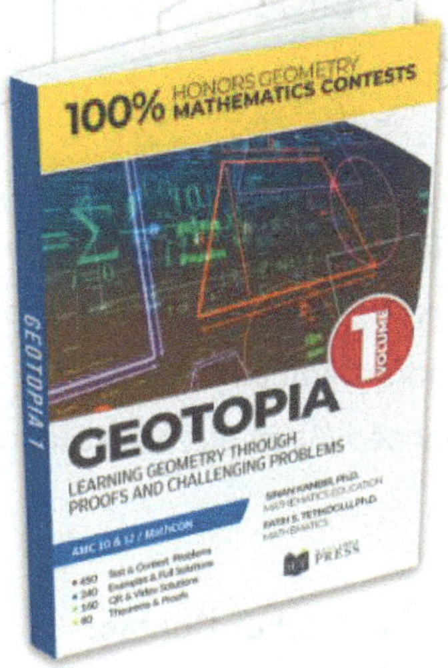

MATHTOPIA PRESS

Paper Book

978-1-7356252-8-7

GeoTopia Series

GeoTopia 0.5 — Essential Geometry Review and Workbook

GeoTopia 1 — Foundations of Geometry

GeoTopia 2 — Honors Geometry and Competition Preparation

GeoTopia 3 — Advanced Geometry (AIME and Olympiad Level)

Publisher Contact

MathTopia Press

United States

Disclaimer

This book is designed for educational purposes to support the learning of geometry, mathematical reasoning, and problem solving. While great care has been taken in preparing this text, the publisher and author assume no responsibility for errors or omissions or for damages resulting from the use of the information contained in this book.

Published by

MathTopia Press

MathTopia Academy LLC

Printed in the United States of America

About the Author — Sinan Kanbir, PhD

Sinan Kanbir is a mathematics educator, author, and competition mentor who focuses on helping students develop strong reasoning and problem-solving skills. He is a professor of mathematics education at the University of Wisconsin–Stevens Point, where he prepares future teachers and designs research-based learning experiences for K–12 students.

Dr. Kanbir is the creator of the GeoTopia series, which bridges classroom geometry with mathematical competitions. Through MathTopia Academy, he teaches advanced courses, summer camps, and enrichment programs that emphasize visual reasoning, elegant proofs, and creative problem solving.

He also contributes problems and curriculum ideas for organizations including MathCON, the Central Wisconsin Math League, and the MAA Wisconsin Section Math Contest.

Dr. Kanbir aims to help students see mathematics as logical, creative, and intellectually beautiful.

Welcome to GeoTopia 2

A Message to Students (GeoTopia 2)

GeoTopia 2 is designed for students who are ready to move beyond foundational geometry and begin thinking like true mathematical problem solvers. Building on the ideas introduced in GeoTopia 1, this book develops deeper reasoning, elegant proof techniques, and advanced problem-solving strategies aligned with high-level mathematical thinking.

Geometry is not about memorizing formulas—it is about seeing structure, recognizing patterns, and constructing logical arguments. In this journey, you will learn to approach problems with insight, creativity, and precision.

In this book, you will explore:

- Circle geometry, cyclic configurations, and power of a point
- Coordinate and analytic geometry methods
- Trigonometric approaches to geometric problems
- 3D geometry and spatial reasoning
- Advanced angle chasing and complex geometric configurations
- Out-of-the-box geometry and creative problem-solving approaches
- Introduction to AIME-level geometric thinking
- Competition-level strategies and problem-solving frameworks

Many problems in this book are inspired by contests such as AMC 10–12, Math League, and MathCON, where success requires not only knowledge, but also creativity, flexibility, and deep insight.

As you work through GeoTopia 2, your goal is not simply to find answers—but to understand why the mathematics works. The ability to explain, justify, and connect ideas is at the heart of true mathematical mastery.

Why Study Advanced Geometry?

Geometry develops some of the most powerful thinking skills in mathematics.

Students who become strong in geometry learn to:

- Visualize complex structures
- Recognize patterns inside diagrams
- Build clear logical arguments
- Solve multi-step problems efficiently

These skills are essential for success in:

- mathematics competitions
- advanced mathematics courses
- engineering and computer science
- scientific reasoning and research

Many AMC and Math League problems reward students who can **interpret diagrams quickly and discover hidden relationships**.

Through carefully chosen examples and challenging practice problems, GeoTopia 2 helps students strengthen these abilities step by step.

By the end of this journey, geometry will become not only a subject—but a powerful **tool for mathematical thinking**.

How This Book Helps You Learn

GeoTopia 2 is organized to guide students from conceptual understanding to **competition-level reasoning**.

Each section includes four important parts:

Key Concepts

Short explanations introduce the geometric ideas needed for the topic.

Worked Examples

Examples demonstrate powerful strategies used to solve challenging geometry problems.

Practice Problems

Problems gradually increase in difficulty, helping students build confidence and fluency.

Challenge Problems

Advanced questions inspired by:

- **AMC 10–12**
- **Math League contests**
- **MathCON advanced levels**

These problems encourage creative thinking and the discovery of elegant solutions.

By solving problems across different levels, students learn how to approach unfamiliar situations with confidence and logical reasoning.

How to Use This Book Successfully

To gain the most from GeoTopia 2, try to develop the habits used by successful mathematicians.

- **Draw clear and accurate diagrams**.

Many geometric insights appear only after carefully analyzing a diagram.

- **Look for relationships.**

Parallel lines, symmetry, similar triangles, and cyclic quadrilaterals often reveal hidden structure.

- **Explain each step of your reasoning.**

Understanding why something works is more important than memorizing formulas.

- **Challenge yourself with difficult problems.**

Growth often happens when solving problems that require multiple ideas.

- **Study elegant solutions.**

Many geometry problems have beautiful and surprising approaches.

Remember

Geometry is not about memorizing formulas.

It is about **seeing structure, reasoning logically, and discovering beautiful mathematical relationships**.

GeoTopia 2 is designed to help you develop the mindset of a **mathematical problem solver**.

Competition Alignment — GeoTopia 2

GeoTopia 2 is designed to bridge **honors geometry coursework and competition mathematics**. The structure of the book reflects the style, rigor, and reasoning required in several major math competitions and academic programs.

Many of the problems and examples are inspired by the types of geometric reasoning used in contests such as:

Competition	Alignment
AMC 10 (Geometry)	✓ Strong alignment
AMC 12 (Geometry)	✓ Moderate–Advanced alignment
Math League Contests	✓ Full alignment
MathCON (Advanced Levels)	✓ Strong alignment
Honors Geometry Courses	✓ Full coverage

AMC 10 / AMC 12

The **American Mathematics Competitions (AMC)** include challenging geometry problems that require creativity, diagram interpretation, and logical reasoning.

GeoTopia 2 prepares students for these problems by emphasizing:

- angle chasing strategies
- similarity and geometric transformations
- circle geometry and cyclic quadrilaterals
- power of a point
- multi-step diagram reasoning

Math League Contests

Math League competitions emphasize **creative problem solving and mathematical insight**. Geometry problems often combine multiple ideas in a single diagram.

GeoTopia 2 helps students develop the flexible thinking required for these problems through carefully designed challenge sets.

MathCON

MathCON problems reward students who understand **visual structure and conceptual reasoning**.

GeoTopia 2 develops these skills through:

- diagram-driven problem solving
- geometric pattern recognition
- elegant and efficient reasoning strategies

Students who complete the problem sets in this book will build strong preparation for **honors courses, contests, and advanced mathematical study**.

Acknowledgments

This book was shaped by a shared love of mathematics and the joy of deep thinking. The author gratefully acknowledges the exceptional problem-solving insight, thoughtful feedback, and careful review provided by the following students:

- **Ayaan Garg** (Massachusetts)
- **Itai Firstenberg** (California)
- **Harika Appalla** (California)
- **Holly Speranza** (Rhode Island)
- **Ayse Lina Saruhan** (California)
- **Musta Alp Saruhan** (California)

Their creativity, precision, and commitment to mathematical clarity meaningfully strengthened this book.

The author also extends sincere thanks to the instructors of **MathTopia Academy—M. Annayev** and **Andrew Carratu (MIT)**—for their mentorship, collaboration, and continued dedication to nurturing mathematical talent.

Finally, heartfelt appreciation goes to the students from **MathPath, MathCON**, and the **MathTopia Academy Summer Math Camps,** whose curiosity, persistence, and enthusiasm for problem solving helped inspire and shape the vision and development of *GeoTopia 2*.

TABLE OF CONTENTS

PRE-TEST

CHAPTER 01

CHAPTER 02

CHAPTER 03

CHAPTER 04

CHAPTER 05

CHAPTER 06

CHAPTER 07

CHAPTER 08

AMC 10 & 12 PRACTICES

SOLUTIONS&Answers

GeoTopia 2 Pre-Background Test & Evaluation

1. In the figure, $m\angle AOC = 16°$ and $m\angle COB = 48°$.

Find the measure of the angle which is formed by the angle bisectors of $\angle AOC$ and $\angle COB$.

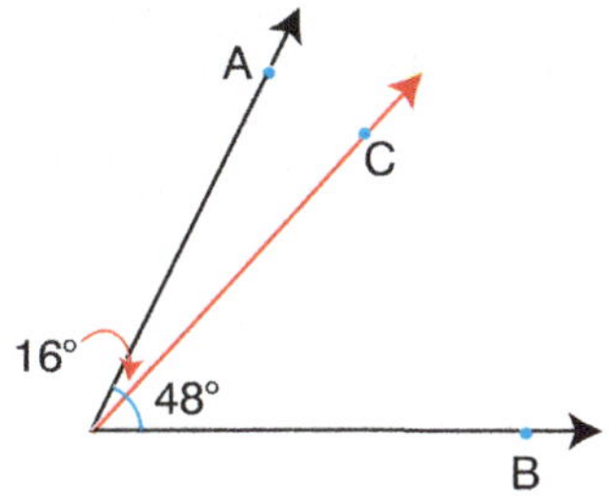

2. Two angles of an isosceles triangle measure 70 and X.

What is the sum of the three possible values of x?

3. In the given figure, ABC is a triangle, The trisections of angles B and C meet at points D and E, and $m\angle BAC = 30°$.

What is $m\angle BDC = \alpha$?

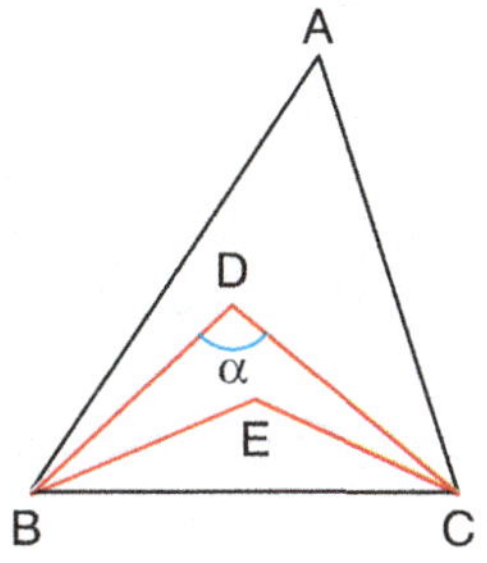

4. ABC is a triangle with $|AB| = |AD| = |BE|$.

$m\angle A = 114°$ and $m\angle B = 60°$. Find $m\angle EDC$.

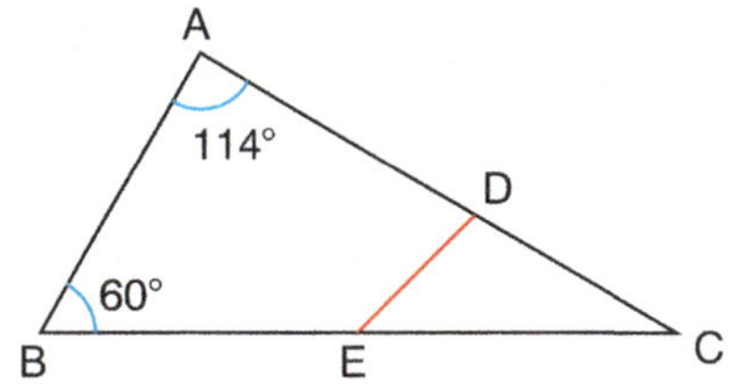

5. In ΔABC, suppose $AB > BC$ and

$m\angle B = 5x - 10$, $m\angle C = x + 25$, $m\angle B + m\angle C < 150°$.

How many integer values of x are possible?

MathTopia Press

6. **(AMC 8, 2024, #8)**

The coordinates of ΔABC are A(5, 7), B(11, 7) and C(3, y), with $y > 7$. The area of ΔABC is 12. What is the value of y?

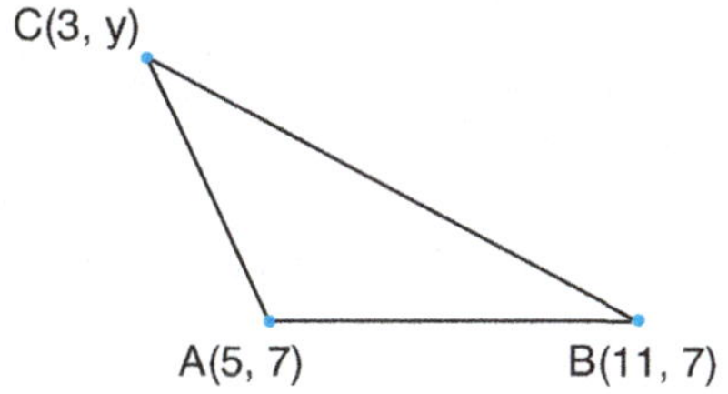

7. ABCD is a quadrilateral with $\overline{AB} \perp \overline{AD}$.

$m\angle ABC = 45°$, $BC = 5\sqrt{2}$, $AD = 10$ and $AB = 17$.

Find DC.

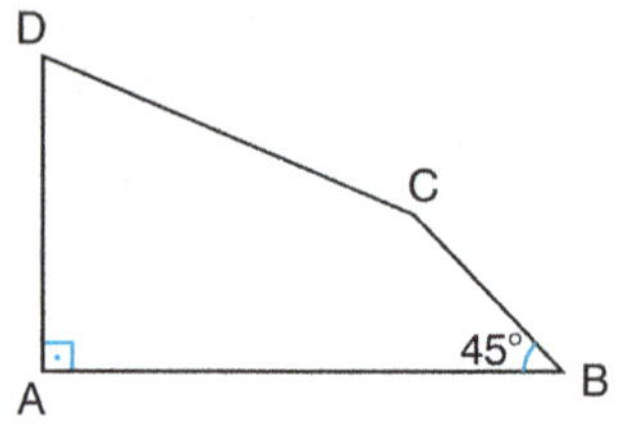

8. ABC is an isosceles right triangle,

$AB = AC$, $BH = 5$ cm, $HC = 3$ cm are given

Find the length $AD = x$.

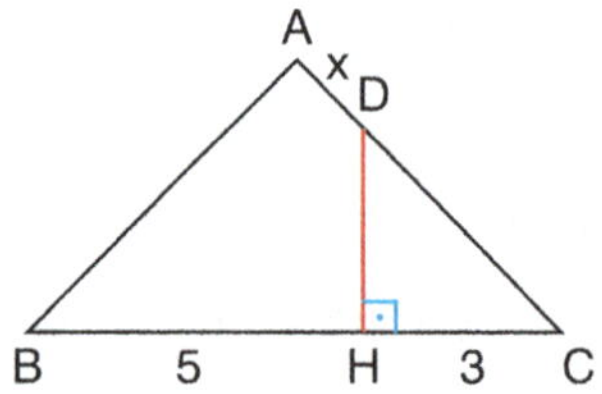

9. In the figure, $BC = 9\sqrt{2}$, $AC = x$, $m\angle B = 75°$.

Find the value of x.

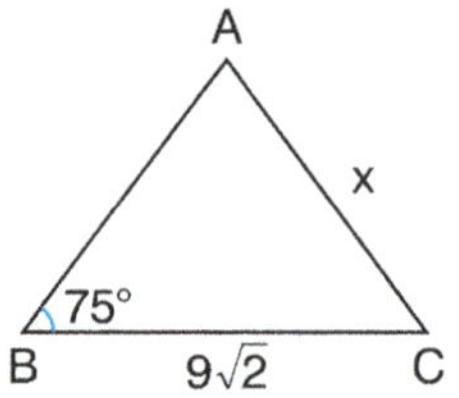

10.

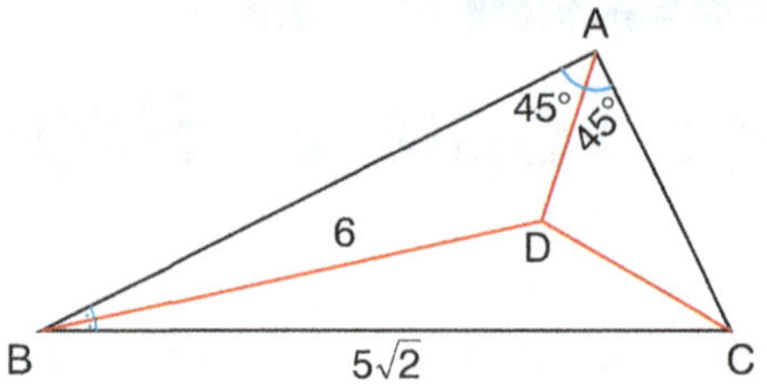

In the figure, BD and AD are angle bisectors of $\angle B$ and $\angle A$, respectively.

$m\angle BAD = m\angle CAD = 45°$, $BD = 6$, and $BC = 5\sqrt{2}$.

Find DC.

11.

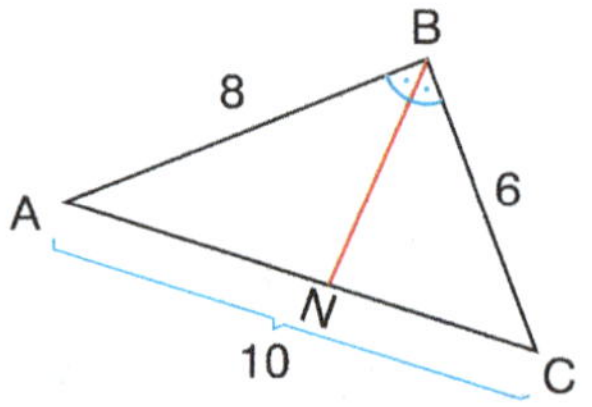

In the figure, $AB = 8$ cm, $BC = 6$ cm, and $AC = 10$ cm.

What is the length of the bisector of $\angle ABC$.

12. In the figure, $\angle DAB = \angle CBD$, $AD = 6$, and $BC = 4$.

What is DC?

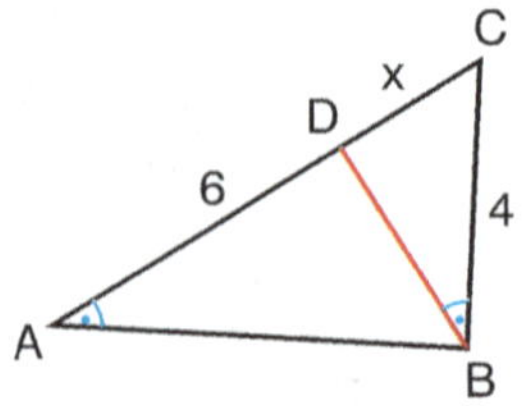

13. In the figure, CB is an angle bisector of $\angle DCA$, $BA \perp AC$, and $DB \perp BC$.

If $AC = 4$, $DC = 9$, and $BC = x$, then find x.

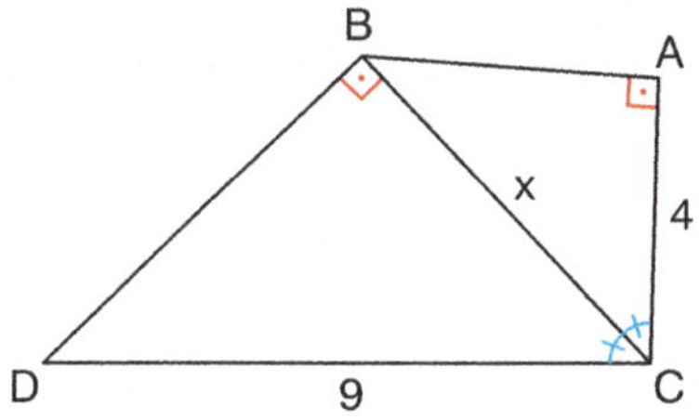

14.

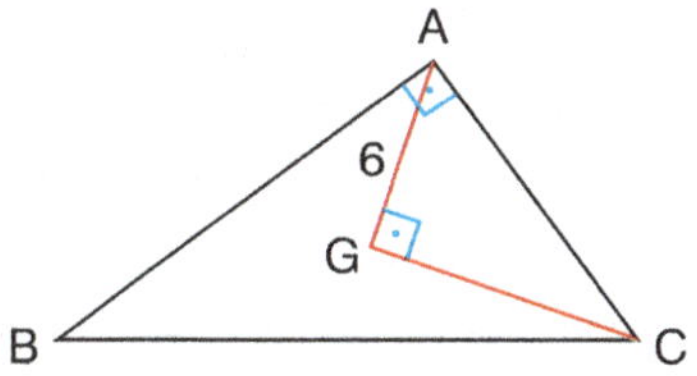

In the figure, point G is the centroid of ΔABC, with $m\angle BAC = 90°$ and $AG \perp GC$.

If $AG = 6$ cm, then find AC.

15. In the figure, DE // BC, and EF is the angle bisector of $m\angle DEC$.

If $AE = 4$ and $DE = EC = 6$, then find the length of BF.

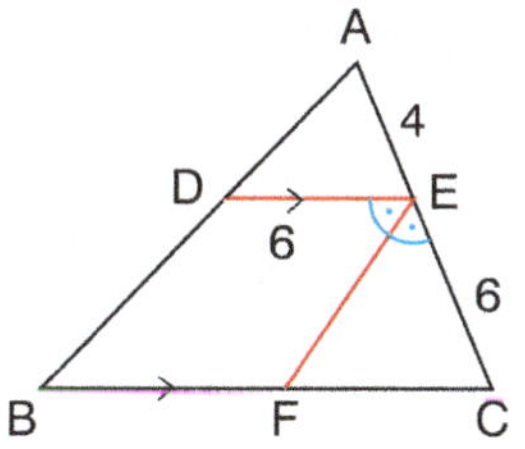

16. **(2017 AMC 8, #16)**

In the figure below, choose point D on $\overline{BC}$ so that ΔACD has perimeters. What is the area of ΔABD?

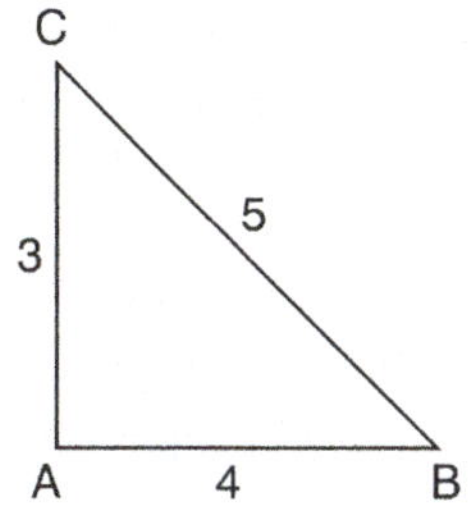

17. In the figure $AB = 5$ cm, $BC = 12$ cm, $CD = 15$ cm, $AD = 14$ cm, with $\angle ABC = 90°$.

The area of a triangle with sides a, b, c is given by Heron's formula: $\Delta = \sqrt{s(s-a)(s-b)(s-c)}$, where $s = \dfrac{a+b+c}{2}$.

Using this, find the area of quadrilateral ABCD.

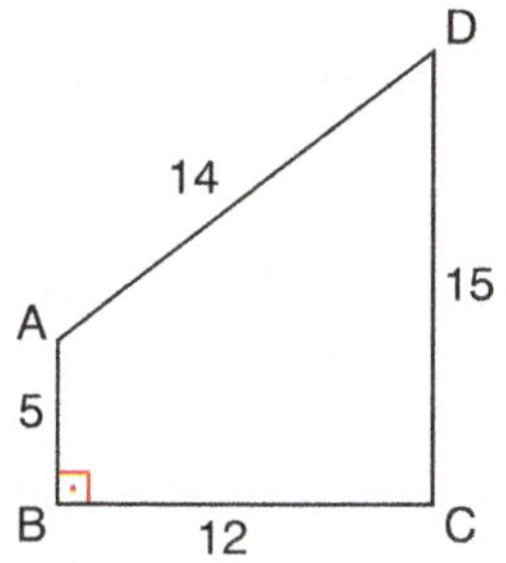

18. Suppose ABC is a right triangle, $\overline{AB} \perp \overline{AC}$ and $\overline{AD} \perp \overline{BD}$. If $AB = AC = 20$ and $BD = 16$, find the area of the triangle BCD.

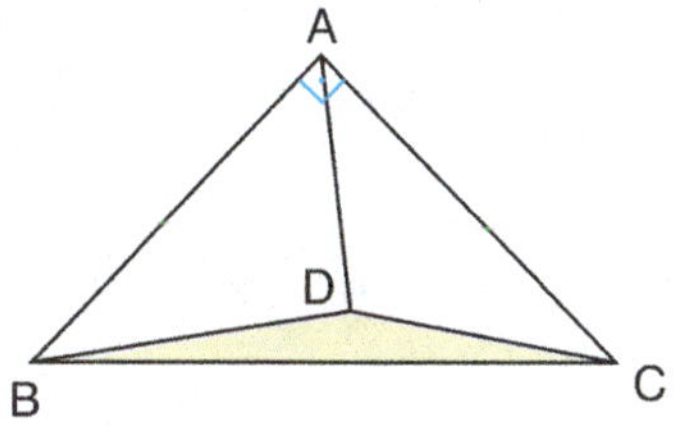

19. In the figure, ABCDE is a regular polygon with $m\angle BDE = 120°$.

How many sides does this polygon have?

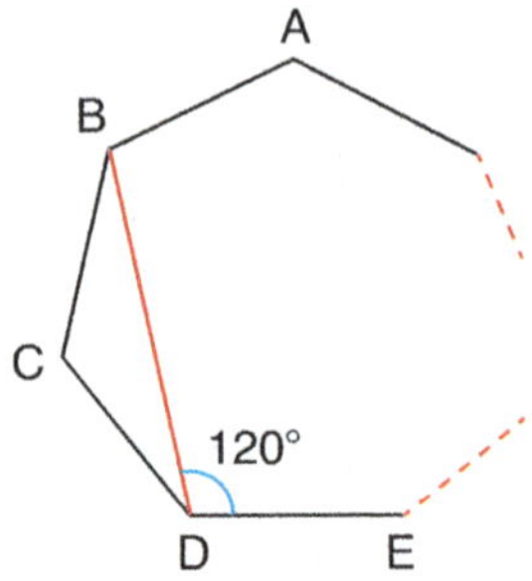

20. In the figure, ABCDEF is a regular hexagon with DH = 3 and HC = 6.

Find AH.

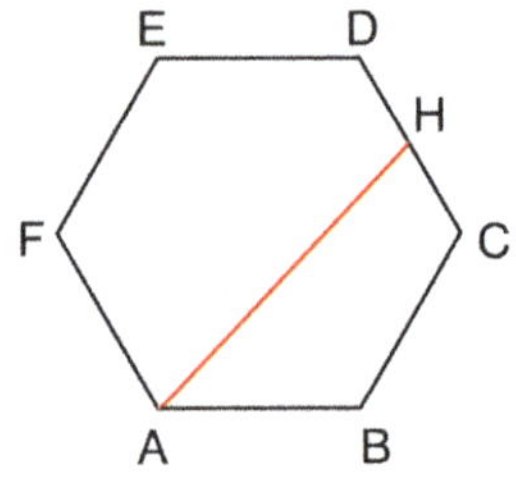

21. In pentagon ABCDE, $\angle CDE = 120°$. The segments satisfy $BC \perp CD$ and $DE \perp EA$, with $|BC| = |EA| = 4$ and $|CD| = |DE| = 3\sqrt{3}$. Find $|AB|$.

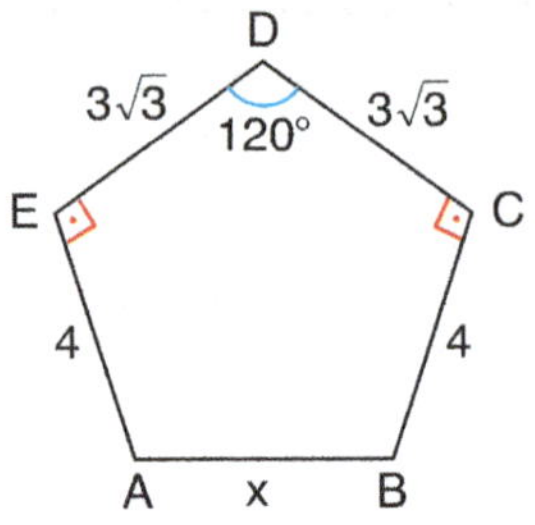

22. **(2003 AMC 8, #21)**

The area of trapezoid ABCD is 164 cm^2. The altitude is 8 cm, AB is 10 cm, and CD is 17 cm. What is BC, in centimeters?

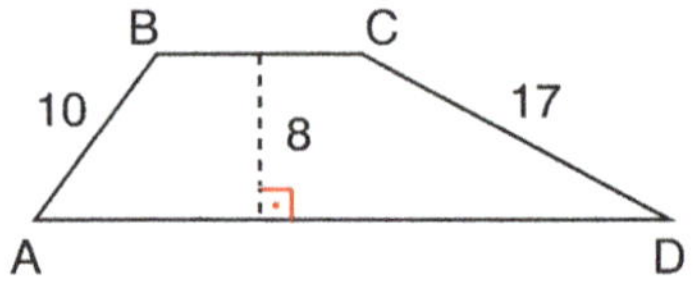

23.

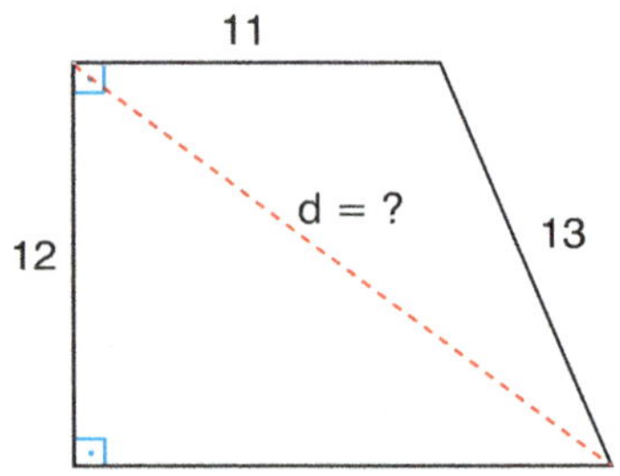

What is the length of the diagonal?

24. Suppose ABCD is a square with AB = 30. If K is the intersection of its diagonals, and L is on $\overline{AB}$ such that AL = 7, what is the value of KL?

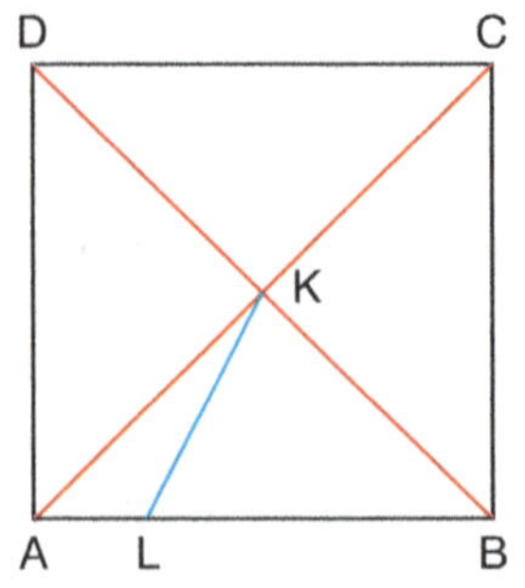

25. ABCD is a parallelogram. E and F are points on $\overline{BC}$ and $\overline{CD}$, respectively, with CE = 3BE, CF = DF, and KF = 6. Find AK.

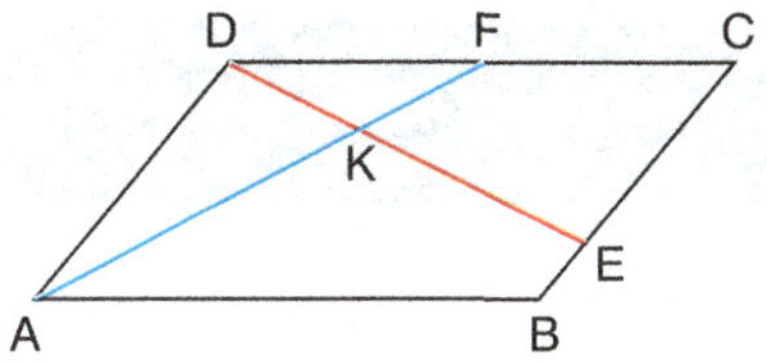

26.

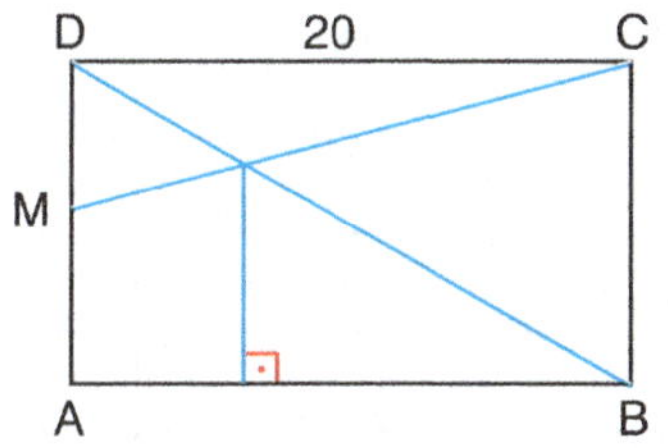

Rectangle ABCD has DC = 20.

Point M lies on AD such that 2MA = 3MD.

Segments intersect at N with NL ⊥ AB and NL = 10.

Find the area of rectangle ABCD.

27. **(AMC8-2008-23)**

In square ABCE, AF = 2FE and CD = 2DE. What is the ratio of the area of ΔBFD to the area of square ABCE?

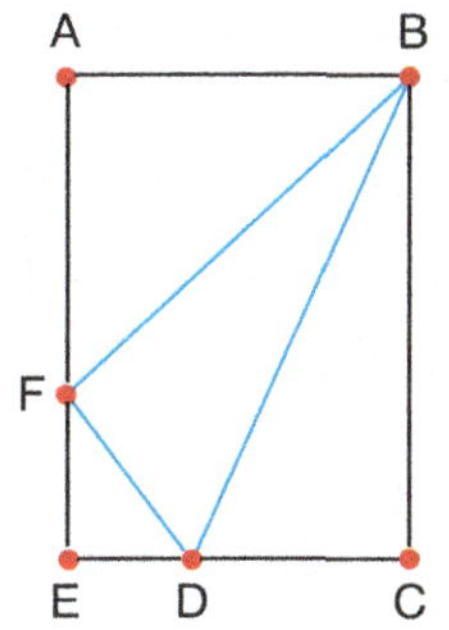

28. A square of area 2 is inscribed in a square of area 3, creating four congruent triangles, as shown below. What is the ratio of the shorter leg to the longer leg in the shaded right triangle?

29. **(2023 AMC 12A, #9)**

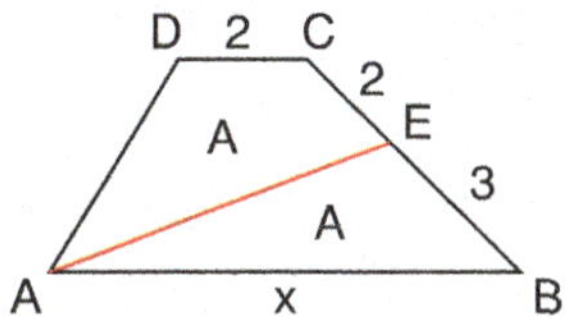

In trapezoid ABCD, AB || CD. Point E is located on BC such that:

- BE = 3,
- EC = 2,
- CD = 2,

If the area of quadrilateral AECD is equal to the area of ΔAEB, find the length of AB, denoted as x.

30. **(2021, AMC 10B, #15)**

In square ABCD, points P and Q lie on $\overline{AD}$ and $\overline{AB}$, respectively. Segments $\overline{BP}$ and $\overline{CQ}$ intersect at right angles at R, with BR = 6 and PR = 7. What is the area of the square?

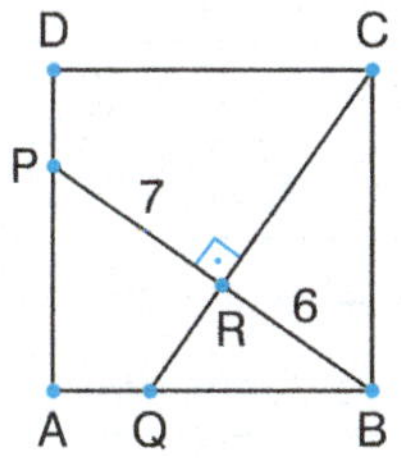

GeoTopia 2 Pre-Background Test Evaluation Table

Problem	Topic	Points Possible	Points Earned	Correct Answer	Notes
1	Angles (Problem 1)	1			
2	Angles (Problem 2)	1			
3	Triangles: Angles & Inequalities (P1)	1			
4	Triangles: Angles & Inequalities (P2)	1			
5	Triangles: Angles & Inequalities (P3)	1			
6	Triangles: Angles & Inequalities (P4)	1			
7	Triangles: Right & Special (P1)	1			
8	Triangles: Right & Special (P2)	1			
9	Triangles: Right & Special (P3)	1			
10	Triangles: Right & Special (P4)	1			
11	Triangles: Lengths (P1)	1			
12	Triangles: Lengths (P2)	1			
13	Triangles: Lengths (P3)	1			
14	Triangles: Lengths (P4)	1			
15	Triangles: Lengths (P5)	1			
16	Triangles: Areas (P1)	1			
17	Triangles: Areas (P2)	1			
18	Triangles: Areas (P3)	1			
19	Polygons: Angles (P1)	1			
20	Polygons: Angles (P2)	1			
21	Polygons: Lengths (P1)	1			
22	Polygons: Lengths (P2)	1			
23	Polygons: Lengths (P3)	1			
24	Polygons: Lengths (P4)	1			
25	Polygons: Lengths (P5)	1			
26	Polygons: Areas (P1)	1			
27	Polygons: Areas (P2)	1			
28	Polygons: Areas (P3)	1			
29	Polygons: Areas (P4)	1			
30	Polygons: Areas (P5)	1			

CHAPTER 1

CIRCLE – ANGLES

Definition 1.1

A **circle** is the set of all points in a plane that are at a fixed distance from a given point in the same plane. The given point is called the **center** of the circle, and the fixed distance is called the **radius**. A radius is also the line segment that connects the center of the circle to any point on the circle. (The plural of radius is **radii**.)

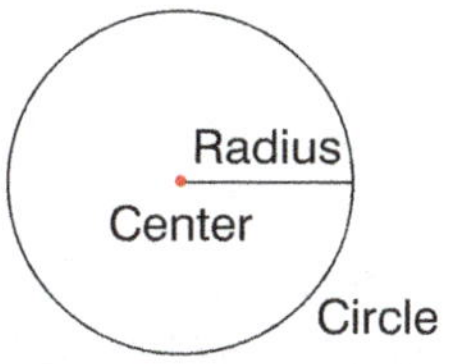

Definition 1.2

A **minor arc** is an arc a circle whose points lie on or between the sides of a central angle. The measure of a minor arc is less than 180°.

Example: Central angle $\angle APB$ determines the minor arc $\overset{\frown}{AB}$.

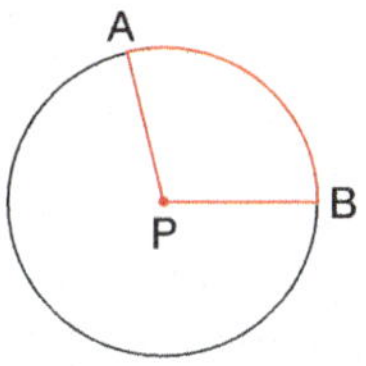

Definition 1.3

Definition - Major Arc

A **major arc** is an arc of a circle whose points lie **on or outside** the sides of a central angle. The measure of a major arc is **greater than 180°**.

Example: Central angle $\angle CQD$ determines the major arc $\overset{\frown}{CD}$.

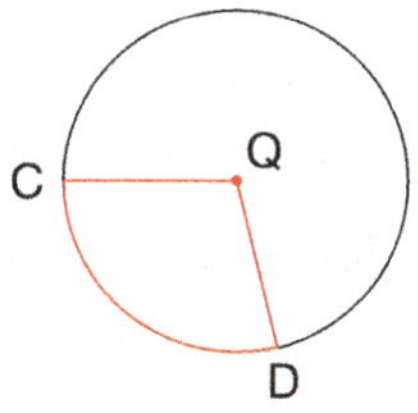

Definition 1.4

Definition - Central Angle

A **central angle** is an angle whose vertex is at the center of a circle and whose sides (rays) intersect the circle at two points. The endpoints of these sides on the circle and the points between them form the **intercepted arc**.

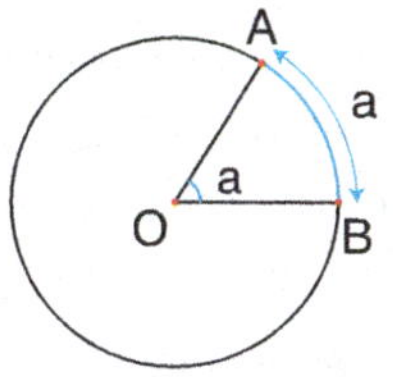

Example: If $\angle AOB$ is a central angle, then arc $\overset{\frown}{AB}$ is its intercepted arc.

Example 1.1

Points A, B, and C lie on a circle with center O.

The measures of the central angles are given as

$m(\angle AOB) = \alpha + 40°$ and $m(\angle ACB) = 4\alpha + 45$.

Find $m(\angle AOB)$.

Solution:

$m(\overset{\frown}{AB}) + m(\overset{\frown}{ACB}) = 360°$

$(\alpha + 40°) + (4\alpha + 45°) = 360°$

$5\alpha = 275° \Rightarrow \alpha = 55°$

$m(\angle AOB) = \alpha + 40° = 95°$

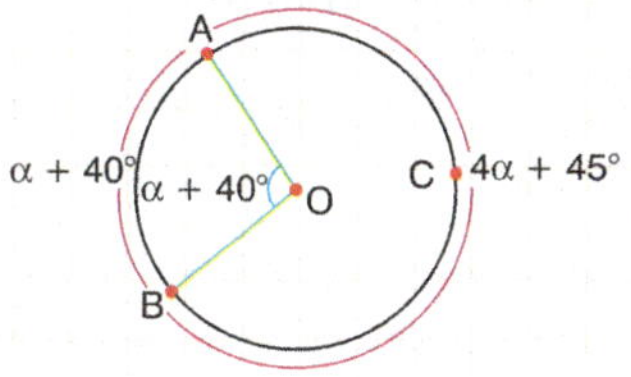

Theorem 1.1

1. Central Angle Theorem

Theorem

The measure of a central angle equals the measure of its intercepted arc. $m(\angle AOB) = m(\overset{\frown}{AB})$

Proof:

A central angle has its vertex at the center of the circle, and its sides pass through the endpoints of the intercepted arc. Since the circle measures 360° and each arc corresponds directly to the central angle subtending it, the measure of the central angle equals the measure of the arc it intercepts.

$$m(\angle AOB) = m(\overset{\frown}{AB})$$

Theorem 1.2

2. Inscribed Angle Theorem

Theorem

The measure of an inscribed angle is half the measure of its intercepted arc. $m(\angle ACB) = \frac{1}{2}m(\overset{\frown}{AB})$

Proof:

Let O be center of the circle. Join O to points A, B, and C. Then $m(\angle AOB)$ is a central angle that intercepts the same arc as $\angle ACB$. From the central angle theorem, $m(\angle AOB) = m(\overset{\frown}{AB})$.

Triangles OAC and OBC are isosceles, so

$m(\angle ACB) = \frac{1}{2}m(AOB)$.

Therefore, $m(ACB) = \frac{1}{2}m(\overset{\frown}{AB})$.

Theorem 1.3

3. Angles Formed by Two Chords Inside a Circle

Theorem

If two chords intersect inside a circle (but not at the center), the measure of each angle formed equals half the sum of the intercepted arcs. $m(\angle 1) = \frac{1}{2}(m(\overset{\frown}{AB}) + m(\overset{\frown}{CD}))$

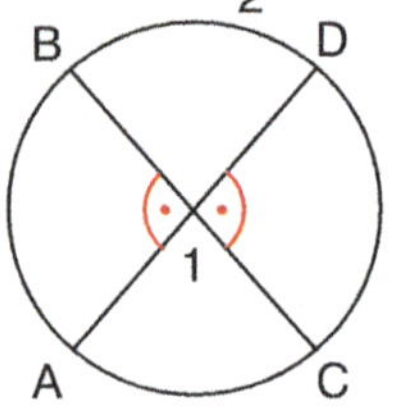

Proof:

Let the two chords AC and BD intersect at E. Draw radii to points A, B, C, and D. Each chord intercepts creates two vertical angles, and each of those angles intercepts two arcs. By combining the inscribed angle relationships,

$m(\angle 1) + m(\angle 2) = \frac{1}{2}m(\overset{\frown}{AB}) + m(\overset{\frown}{CD})) + \frac{1}{2}(m(\overset{\frown}{BA}) + m(\overset{\frown}{DC})) = 180°$.

Thus, for one of them, $m(\angle 1) = \frac{1}{2}(m(\overset{\frown}{AB}) + m(\overset{\frown}{CD}))$.

Theorem 1.4

4. Angles Formed Outside the Circle

Theorem

If a tangent and secant, two tangents, or two secants intersect outside a circle, the measure of the angle formed equals half the difference of the intercepted arcs.

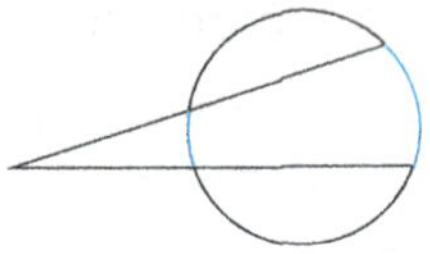

$m(\angle 1) = \frac{1}{2}(m(\overset{\frown}{large}) - m(\overset{\frown}{small}))$

Proof:

Let two secants intersect outside the circle at point P. Draw the intercepted arcs $\overset{\frown}{AB}$ (large arc) and $\overset{\frown}{CD}$ (small arc). Construct auxiliary radii to show that the external angle and the intercepted arc together from a full circle. By subtracting the smaller intercepted arc from the larger and halving, we obtain:

$$m(\angle 1) = \frac{1}{2}(m(\overset{\frown}{large}) - m(\overset{\frown}{small})).$$

Theorem 1.5

5. Tangent - Chord Angle Theorem

Theorem

An angle formed by a tangent and chord through the point of tangency is equal to half the measure of its intercepted arc. $m(\angle 1) = \frac{1}{2}(m(\overset{\frown}{AB})$

Proof:

The tangent at point A is perpendicular to the radius OA. Join chord AB. The central angle $\measuredangle AOB$ intercepts the same arc as $\angle 1$. Since $\angle 1$ is an inscribed angle (vertex on the circle), it measures half the intercepted arc:

$$m(\angle 1) = \frac{1}{2}(m(\overset{\frown}{AB})$$

Example 1.2

The figure shows a semicircle with center O. Find $m(\angle BED) = x$.

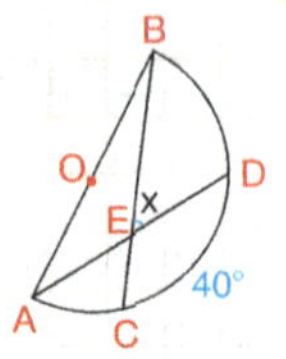

Solution:

If we complete the circle we can write

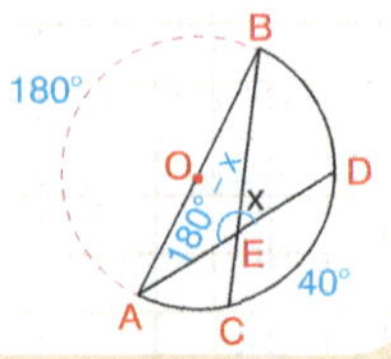

$180° - x = \dfrac{180° + 40°}{2}$

$180° - x = 110°$

$x = 70°$

Example 1.3

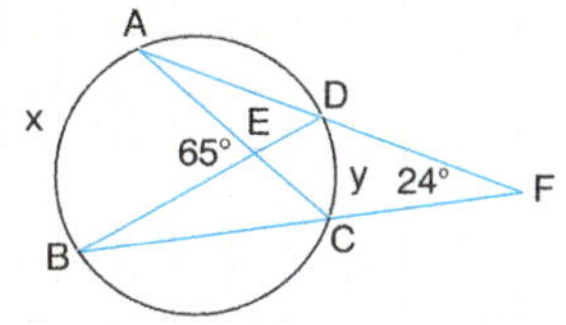

Solution:

Let $m(\widehat{AB}) = x$ and $m(\widehat{CD}) = y$.

$\frac{1}{2}(x + y) = 65$, $\frac{1}{2}(x - y) = 24$

$x + y = 130$, $x - y = 48$

$2x = 178 \Rightarrow x = 89$, $y = 41$.

Example 1.4

By the inscribed angle theorem:

$m(\widehat{BC}) = 2x$, $m(\widehat{DC}) = 4x$.

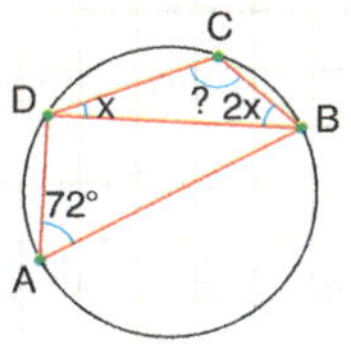

Solution:

Since $m(\angle DAB) = 72°$:

$\frac{1}{2}m(\widehat{DCB}) = 72° \Rightarrow$

$\frac{1}{2}(6x) = 72° \Rightarrow x = 24°$.

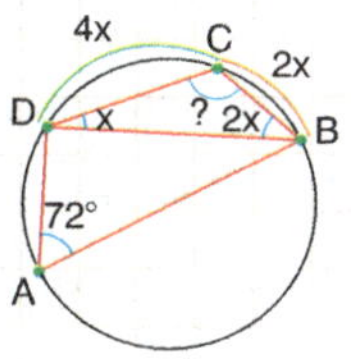

Then:

$m(\angle C) + 3x = 180° \Rightarrow m(\angle C) + 72°$

$= 180° \Rightarrow m(\angle C) = 108°$.

Example 1.5

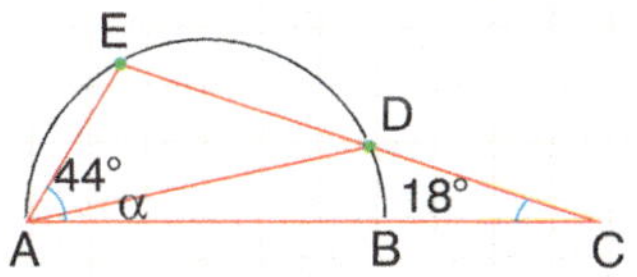

AB is the diameter of the semicircle shown.

Find $m(\angle DAB) = \alpha$.

Solution:

Let the arc AE = x and the arc EB = y.

From the semicircle,

$x + 88 + y = 180$.

From the tangent–secant angle at C:

$x - y = 36$.

Solving (1) and (2): $x = 64°$, $y = 28°$.

Since $\alpha = \frac{1}{2}y$, $\alpha = 14°$.

Property 1.1

Angle Formed Outside a Circle by Two Tangents

Let PA and PB be tangents to a circle with center, O, touching the circle at points A and B.

$$m(\angle APB) = x,\ m(\widehat{ACB}) = \alpha,\ m(\widehat{AB}) = \beta$$

Then the measure of the external angle formed by the two tangents is:

$$x = \frac{\alpha - \beta}{2}$$

and

$$x + \beta = 180°$$

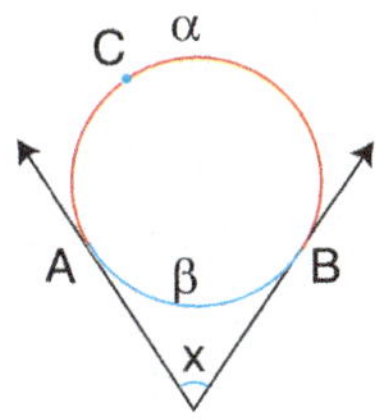

Example 1.6

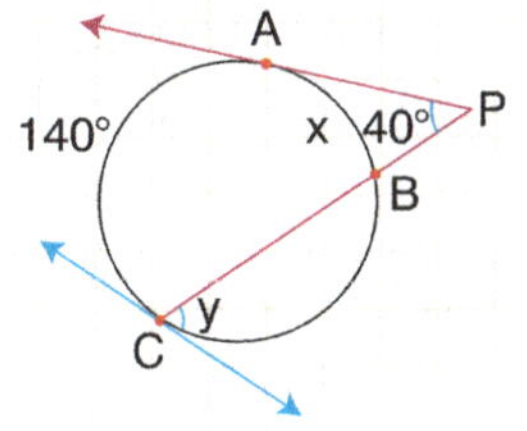

Solution:

PA is tangent A. The major arc through C is 140°. The exterior angle at P is 40°.

1. **Use exterior angle theorem (tangent-secant):**

   $m\angle APB = \frac{1}{2}(m(\widehat{ACB}) - m(\widehat{AB}))$

   $40 = \frac{1}{2}(140 - x) \Rightarrow x = 60°$.

2. **Use inscribed angle theorem:** an inscribed angle equals half its intercepted arc.

3. **Arc sum is 360°:**

   $140 + x + 2y = 360 \Rightarrow 140 + 60 + 2y = 360 \Rightarrow$

   $2y = 160 \Rightarrow y = 80°$.

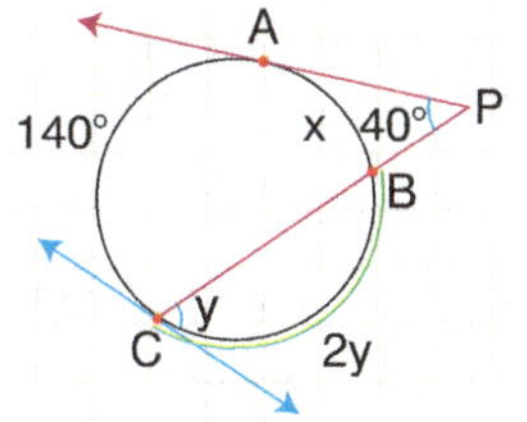

Example 1.7

A line segment AB is the diameter of a semicircle centered at O. Points C and D are located counterclockwise on the semicircle such that m(∠BCD) = 110°. Find m(∠BOD).

Solution:

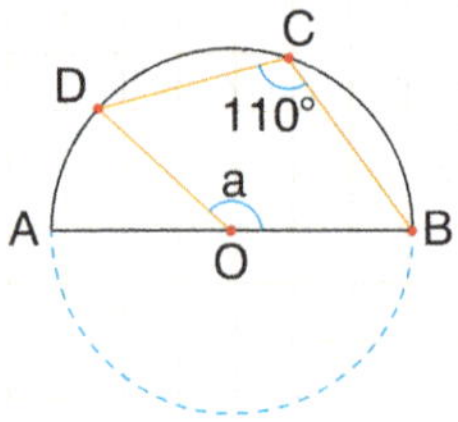

1. From the **inscribed angle theorem,**

   $m(\widehat{BAD}) = 2 \times 110° = 220°$.

2. The arcs of a circle sum to 360°:

   $m(\widehat{BCD}) + 220° = 360° \Rightarrow m(\widehat{BCD}) = 140°$.

3. By the **central angle theorem,**

   $m(\angle BOD) = m(\widehat{BCD}) = 140°$.

Definition 1.5

A **cyclic quadrilateral** is a quadrilateral that can be inscribed in a circle - that is, all four vertices lie on the same circle.

In the figure, ABCD is a cyclic quadrilateral

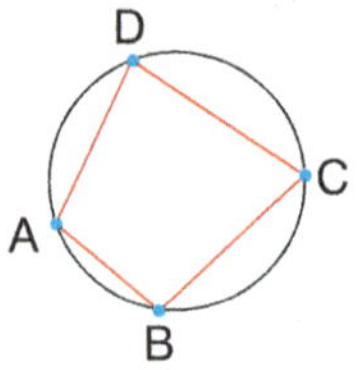

Theorem 1.6

Theorem-Opposite Angles in a Cyclic Quadrilateral

Theorem

In a cyclic quadrilateral, opposite angles are supplementary.

$m(\angle A) + m(\angle C) = 180°$, $m(\angle B) + m(\angle D) = 180°$

Proof:

Let $m(\angle C) = \alpha$ and $m(\angle A) = \beta$.

Since both ∠C and ∠A are inscribed angles, their intercepted arcs are:

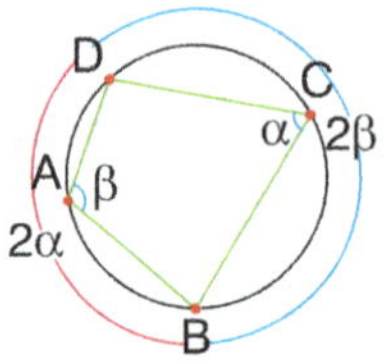

$m(\widehat{BD}) = 2\alpha$, $m(\widehat{BC}) = 2\beta$.

These arcs together form the entire circle:

$2\alpha + 2\beta = 360° \Rightarrow \alpha + \beta = 180°$.

Thus, the opposite angle ∠A and ∠C are supplementary. By the same reasoning, ∠B and ∠D are also supplementary.

Example 1.8

In the figure, AB ⊥ BC, DE ⊥ AC,

m(∠AFB) = 80° and

m(∠EBC) = 20°.

Find m(∠C) = α.

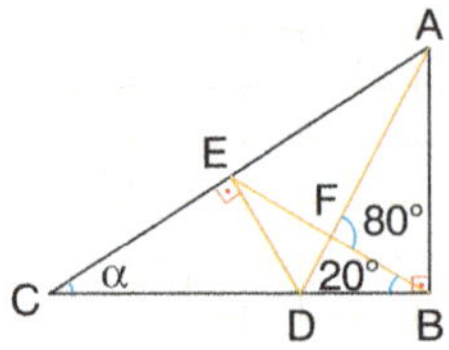

Solution:

1. Since ∠AED = ∠ABD = 90°, the two angles are **supplementary**, so A, B, D, E are concyclic (they lie on the same circle).

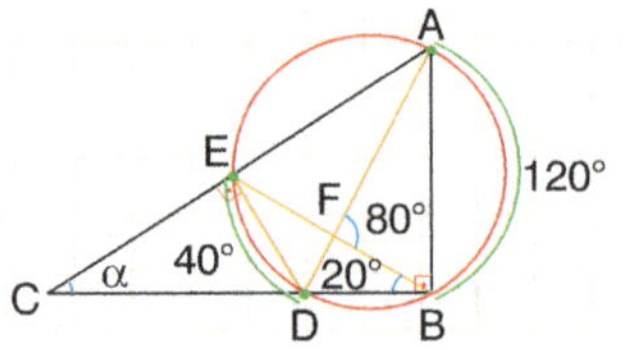

2. In the cyclic quadrilateral ABDE:

   $m(\angle EBD) = 20° = \frac{1}{2}m(\widehat{DE}) \Rightarrow m(\widehat{DE}) = 40°$

3. Using ∠AFB = 80° and the **intersecting chords theorem:**

   $80° = \frac{1}{2}(m(\widehat{AB}) + m(\widehat{DE})) \Rightarrow 80° = \frac{1}{2}(m(\widehat{AB}) + 40°)$

   Hence, $m(\widehat{AB}) = 120°$.

4. Finally, applying the **exterior angle formula** for the circle:

   $\alpha = \frac{1}{2}(m(\widehat{AB}) - m(\widehat{DE})) = \frac{1}{2}(120° - 40°) = 40°$.

Important 1.1

1. Tangent Segments from an External Point

From a point outside a circle, **two tangent segments** can be drawn to the circle. These tangent segments are **equal in length**.

Example:

If PA and PB are tangents from point P to a circle, then PA = PB.

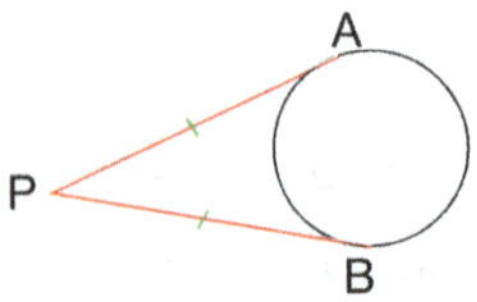

Important 1.2

2. Tangent - Radius Relationship

A tangent line to a circle is **perpendicular** to the radius drawn to the point of tangency. Conversely, if a line through a point on a circle is **perpendicular** to the radius at that point, the line is **tangent** to the circle.

Example:

If BC is tangent to circle O at B, then BC ⊥ OB.

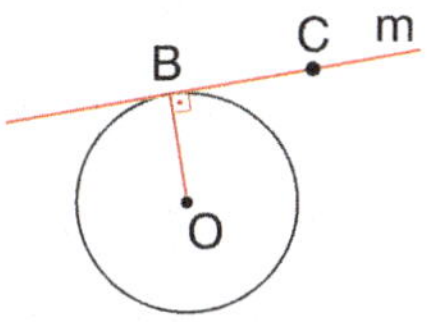

Practice 2

Adapted from a geometric puzzle originally shared by Catriona Agg (@Cshearer41) on X

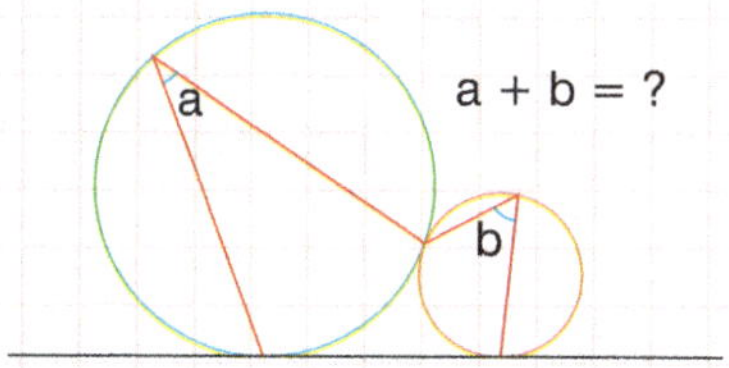

Solution:

By the tangent-chord theorem, each angle (a and b) equals the angle made by the chord and its tangent.

These two angles are complementary, so a + b = 90°.

Practice 3

Adapted from a geometric puzzle originally shared by Catriona Agg (@Cshearer41) on X

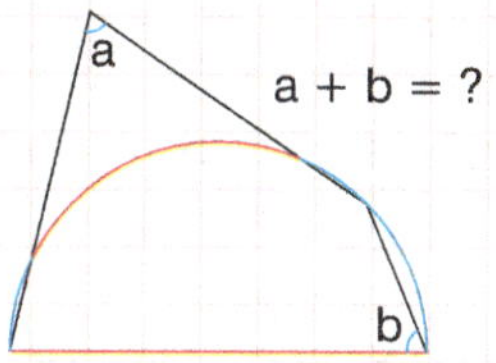

Solution:

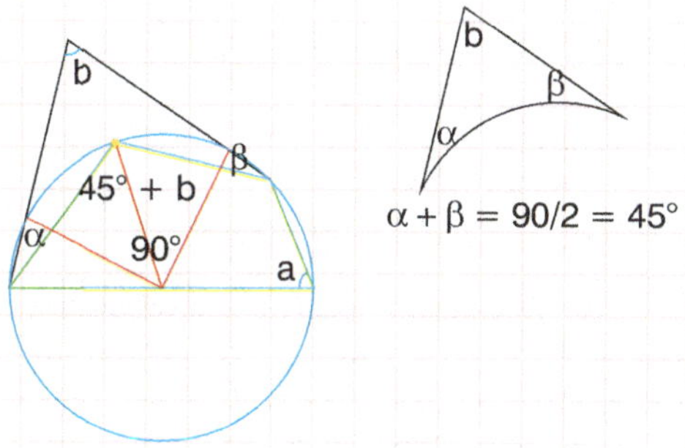

a + 45 + b = 180 so a + b = 180 – 45

a + b = 135°

(opposite angles of cyclic quadrilateral)

Practice 1

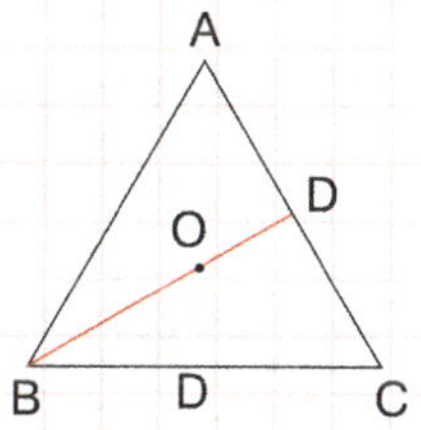

In triangle ABC, m(∠A) = 58° and O is the circumcenter. If BO meets AC at D, find m(∠DBC).

Solution:

Since O is the circumcenter, BO is a radius of the circumcircle. The angle between a radius and the chord through that point equals 90° – the inscribed angle subtending the same chord.

Here, chord BC subtends ∠A.

∠DBC = 90° – ∠A = 90° – 58° = 32°.

Practice 4

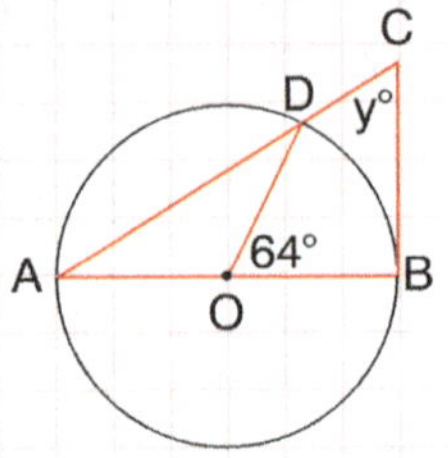

In the figure above, $\overline{BC}$ is tangent to a circle with center O and point D lies on $\overline{AC}$. What is the value of y?

Work Space

Practice 5

Given that the three lines are tangents, find the value of x + y in the figure.

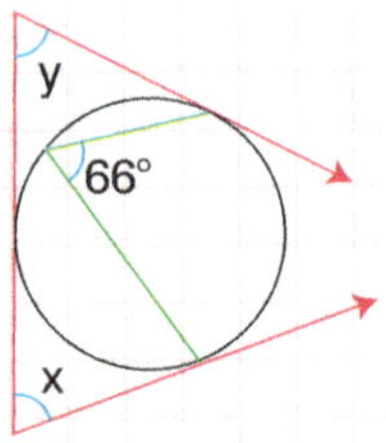

Work Space

Practice 6

Adapted from a geometric puzzle originally shared by Catriona Agg (@Cshearer41) on X

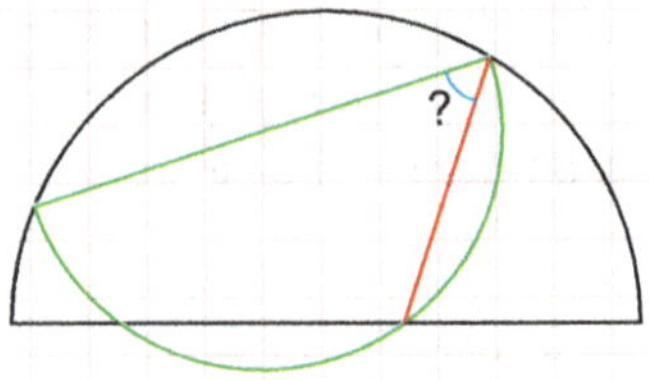

Work Space

Practice 7

Adapted from a geometric puzzle originally shared by Catriona Agg (@Cshearer41) on X

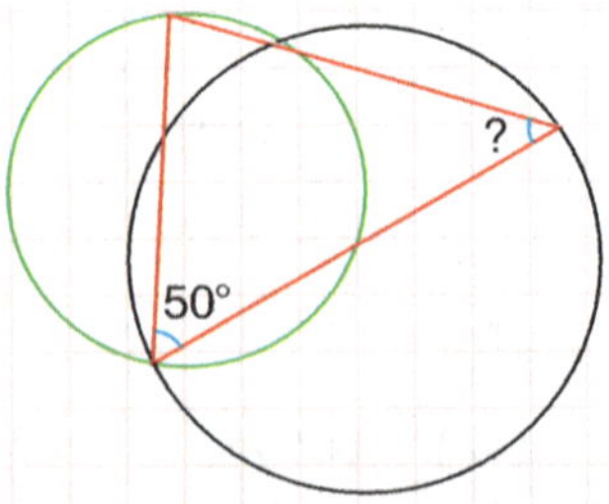

Work Space

1.

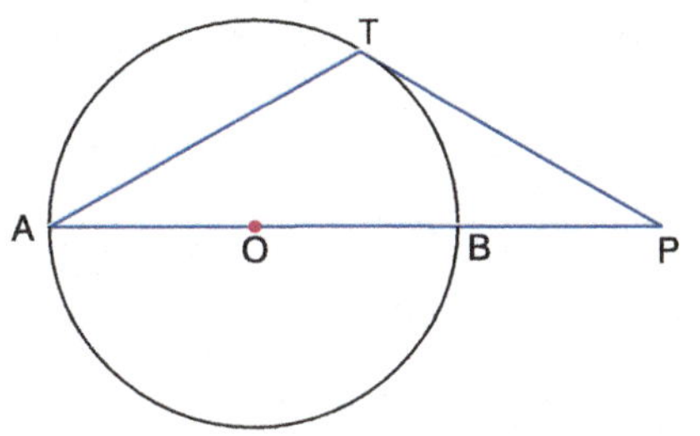

In circle O, the tangent intercepts the circle at T and the diameter at A and B.

If AT = TP, find m $\angle TAP$.

4.

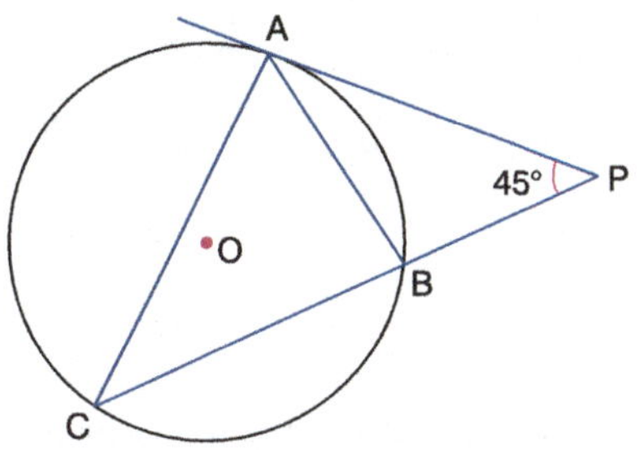

In circle O, the tangent intercepts the circle at A and secant segment PBC are drawn to circle O.

If m $\angle APC = 45°$ and m $\widehat{AC}$: m $\widehat{AB}$ = 5 : 2, find m $\angle CAB$?

2.

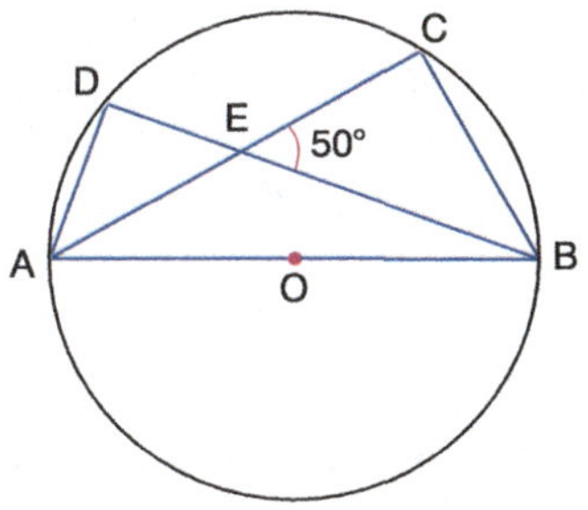

In circle O, AB is a diameter and points C and D are on the circle.

If m $\angle CEB = 50°$, find m $\angle DAC$ + m $\angle DBC$.

5.

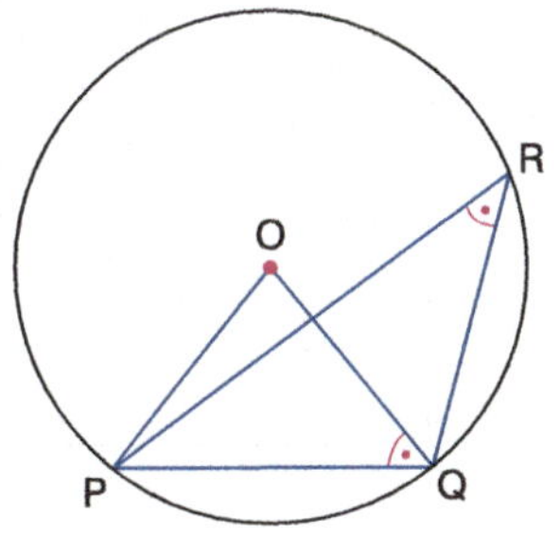

O is the center of the circle in the figure.

Find m $\angle OQP$ + m $\angle PRQ$.

3.

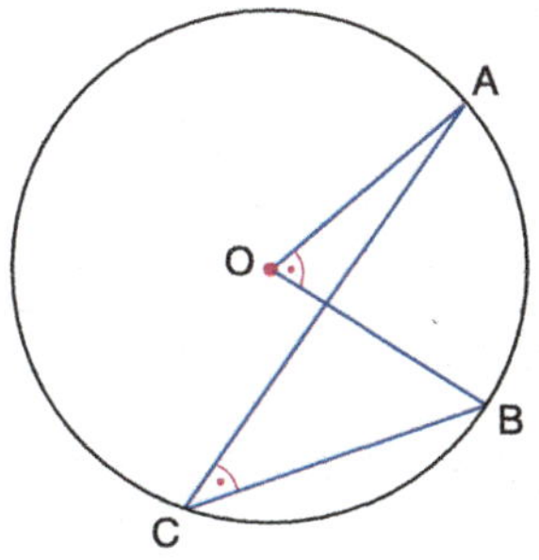

In circle O, m $\angle AOB = (x + 20°)$ and m $\angle ACB = x$.

What is the value of x?

6.

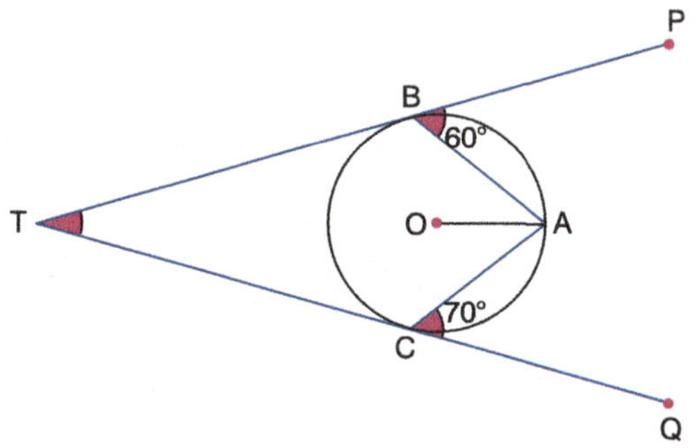

In the figure, TP and TQ are tangent to the circle with center O at B and C, respectively.

If $\angle PBA = 60°$ and $\angle ACQ = 70°$, what is $\angle BTC$ in degrees?

7.

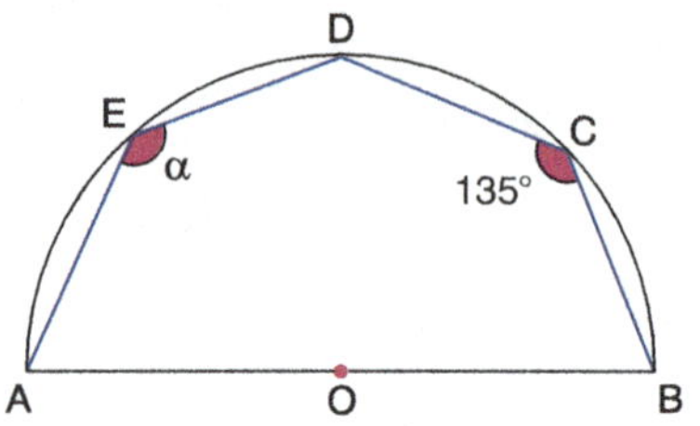

In the figure, O is the center of the semi circle.

If m ∠BCD = 135° and AE = ED = DC = CB, find m ∠AED.

8.

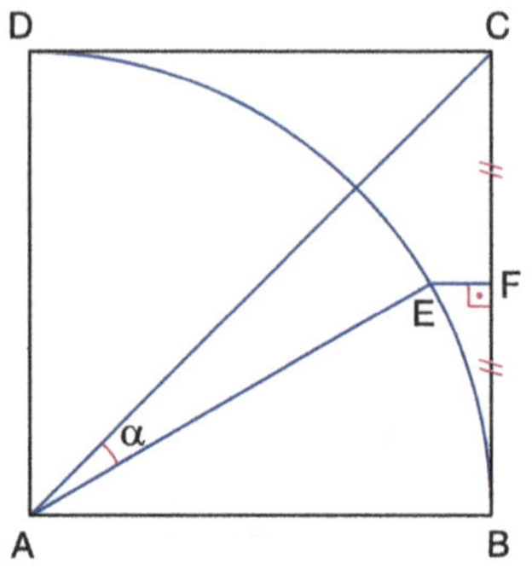

ABCD is a squre and A is the center of the quarter circle.

Given that EF ⊥ BC and CF = FB, find m ∠CAE.

9.

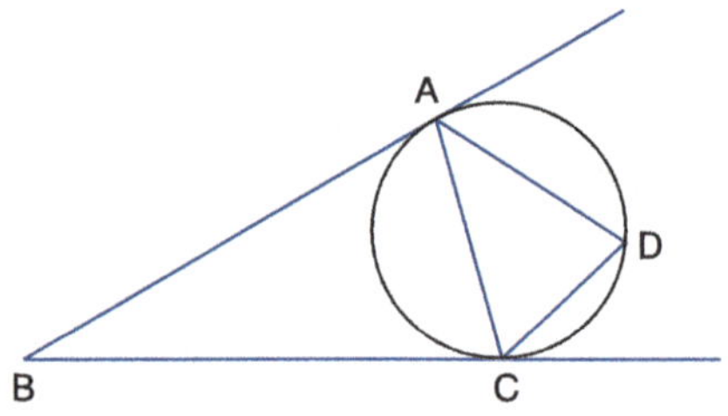

In the diagram above, lines AB and BC are tangent lines of a circle, with points A and C as their respective tangents points.

If ∠ADC = 70°, what is ∠ABC in degrees?

10.

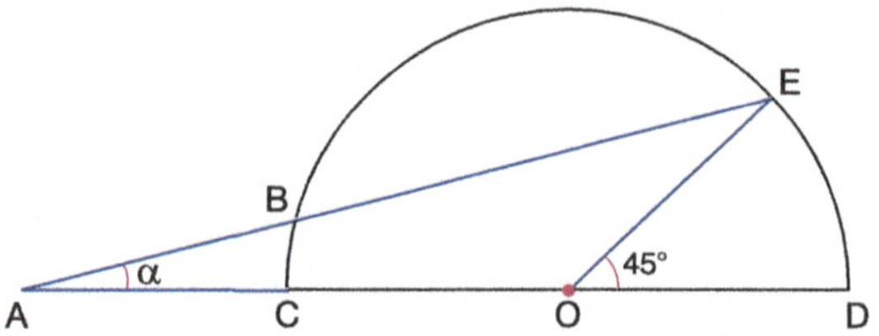

CD is a diameter AB = OD

m ∠EOD = 45°, find m ∠EAD = α

11.

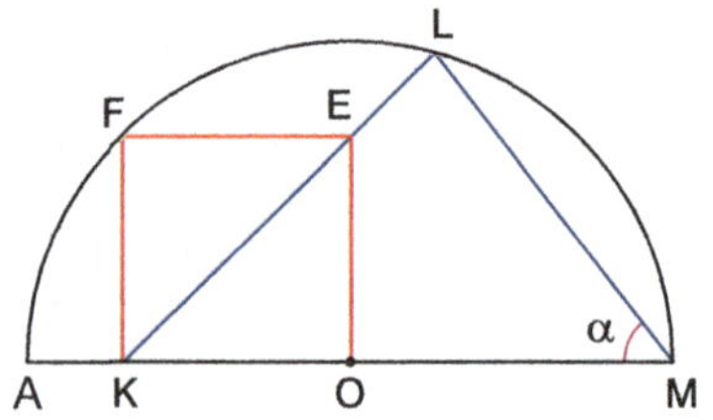

FKOE is a square and AM is a diameter. O is the center of the circle.

What is m ∠KML?

12.

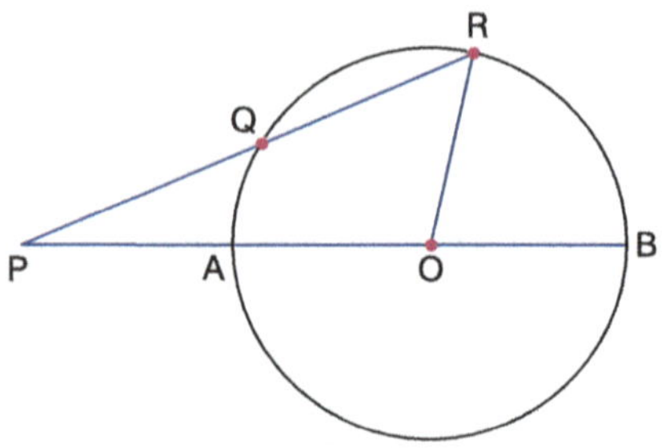

In circle O, AB = 2PQ and m ∠P = 24°

Find m ∠RQB.

1.

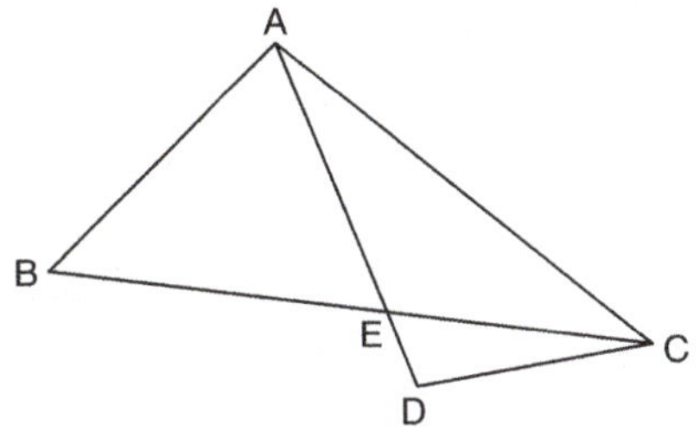

In the figure AB = AD = AC Given that m∠BAD = 48°, find m∠BCD.

2.

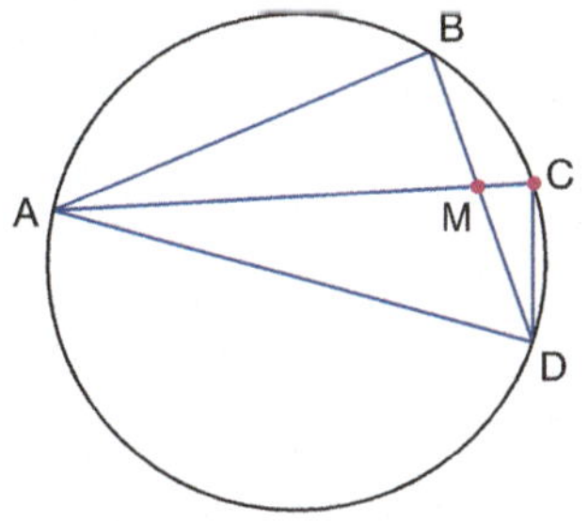

In the figure m∠AMD = 120° and m∠DBA = 90°, and |AB| = 6.

What is AM?

3. Circle O, shown here with chords AB and BE, has secants AC and DE that intersect at X. If m∠ABE = 35° and m∠AXE = 15°, what is the measure of $\overset{\frown}{CD}$?

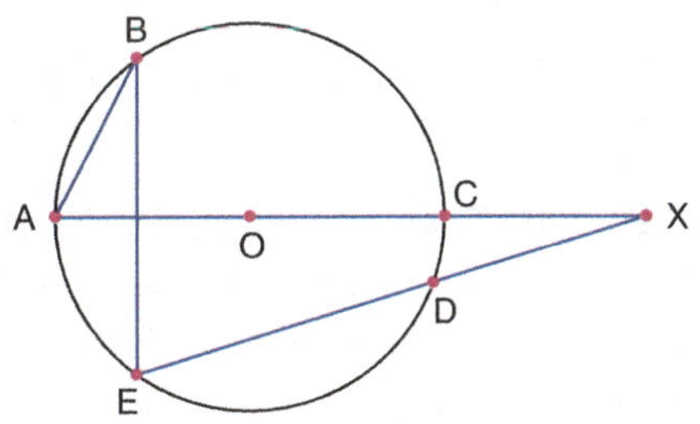

4.

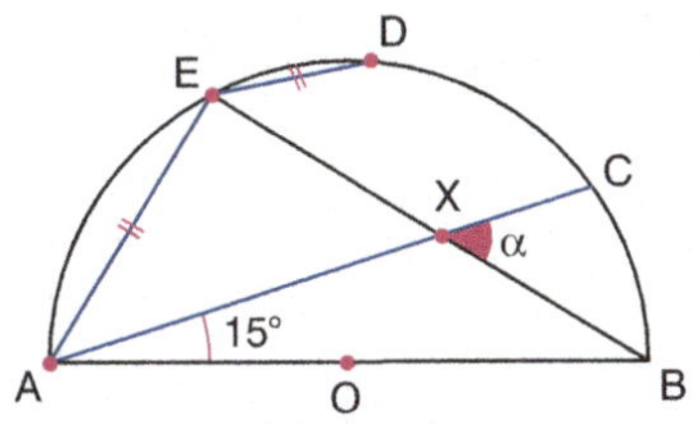

In the figure, AB is a diameter of the semicircle where AC//ED. Given that |AE| = |ED| and m∠CAB = 15° find m∠CXB.

5. Two circles are externally tangent at point C. A line also externally tangent to both circles points A and B. Given that m∠ABC = 48°, find m∠AMC.

6.

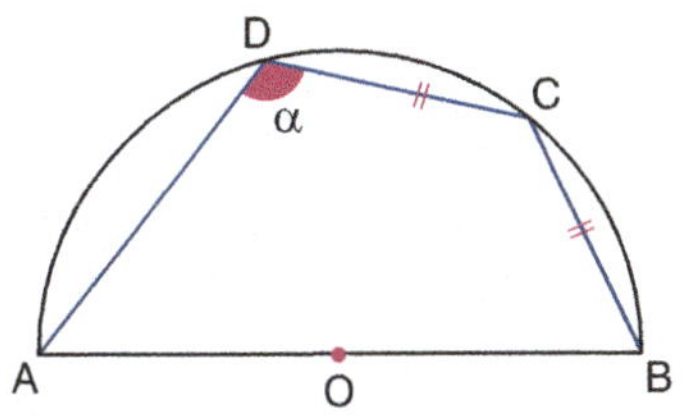

In the figure, AB is the diameter of the semicircle. If m∠BCD = 125° and BC = CD, then find the value of m∠ADC.

7.

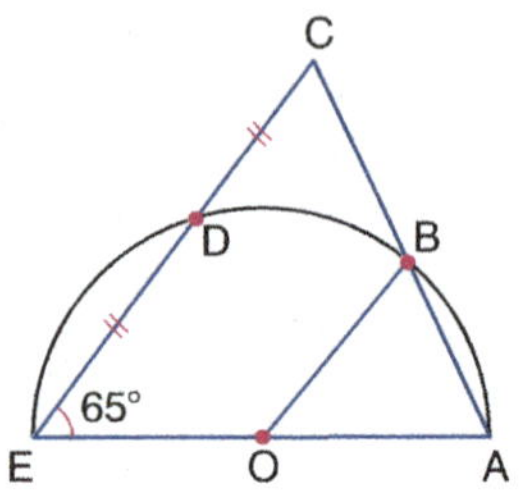

In the figure, O is the center of the semicircle.

If ED = DC and m∠AED = 65°, find m∠ADB.

10.

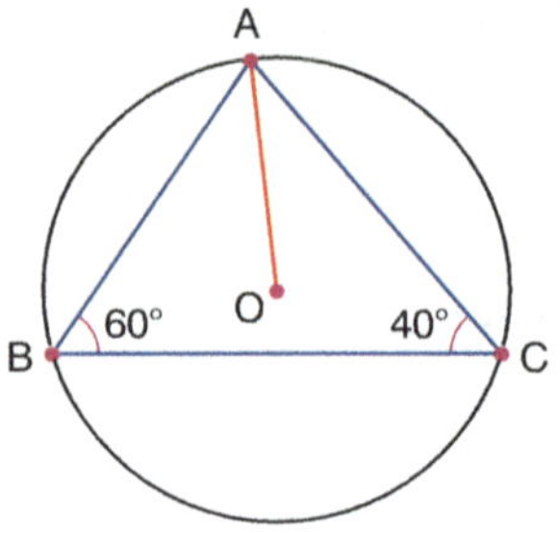

The vertices of the triangle ABC are on the circle with center O. If m∠ABC = 60° and m∠ACB = 40°, find m∠BOA

8.

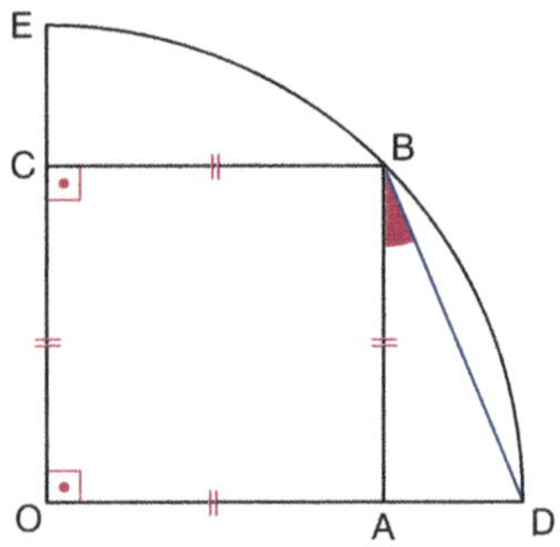

In the figure, OABC is a square and O is a center of $\frac{1}{4}$ circle.

What is m∠ABD?

11.

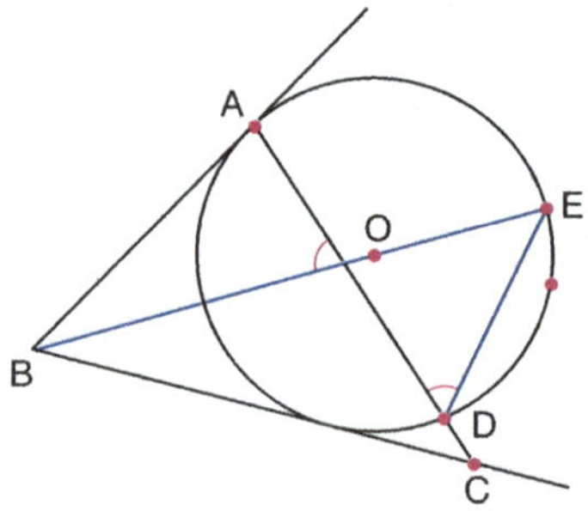

In the figure, O is the center of the Circle and BC = AC.

If m∠ACB = 40°, find the value of m∠ADE.

9. Circle O is tangent to the segment AC and BC at A and B, respectively.

If 3m∠BCA = 2m∠AOB find the measure of ∠BCA.

12. A, B, C, D, E and G are located on a circle clockwise and point P is an outside of the circle such that P, D, E, and P, G, B are collinear. Given that m∠BAE = 130°, m∠ABC = 120° and m$\widehat{CD}$ = 60°, find m∠AED.

1. **(CWML 10-11 Meet 3, #10)**

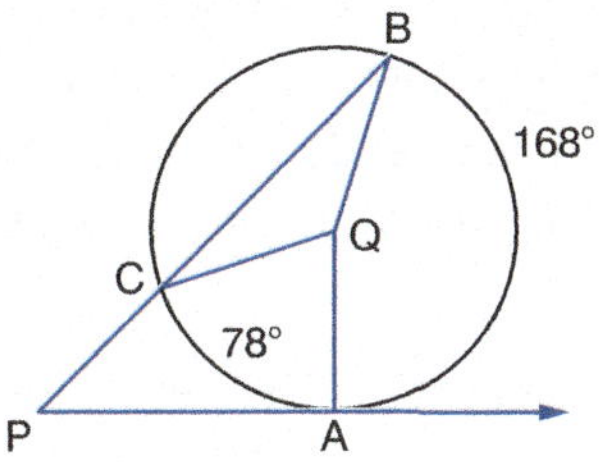

Ray PA is tangent to circle Q at A, arc AC measures 78°, and arc AB measures 168° as shown. The line containing chord CB passes through P. Find the measure of angle P.

2. **(MathCON 19 Final GR10, #12)**

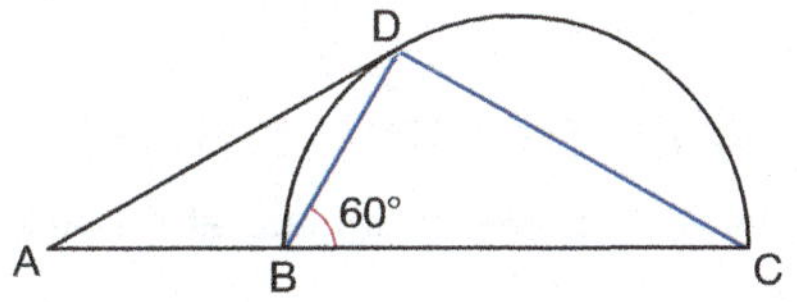

Suppose D is the point of tangency between a semi-circle and a line.

If m DBC = 60°, what is the value of $\frac{BC}{DC}$?

A) $\sqrt{2}$ B) $\frac{3}{2}$ C) $\frac{2\sqrt{3}}{3}$

D) $\frac{4}{3}$ E) None of the preceding

3. **(MathCON 17 Final GR11, #17)**

In the circle below, m(∠ABC) = 50° and m(KL) = 40°.

What is the measurement of the angle m(∠AFC)?

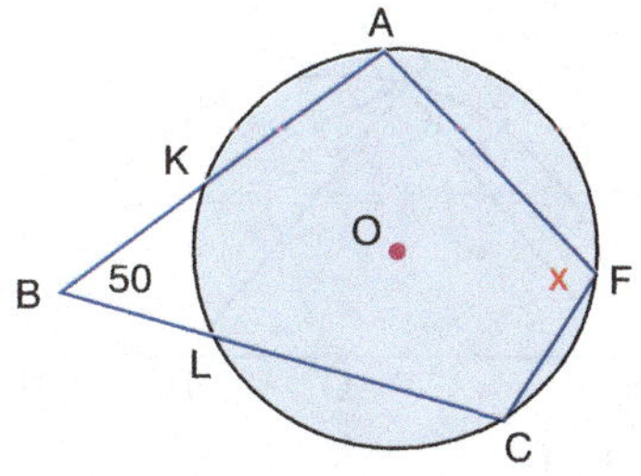

A) 110° B) 115° C) 120° D) 125° E) 135°

4. **(MathCON 21 Final GR12, #6)**

In the figure, points A, B, C, D, and E are on the circle.

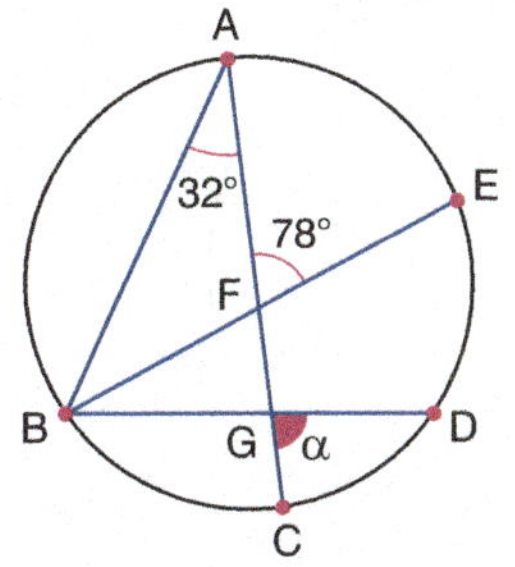

$m\widehat{AE} = 2 \cdot m\widehat{ED}$.

m∠BAC = 32°, and

m∠AFE = 78°.

What is m∠CGD = α?

A) 55° B) 69° C) 73° D) 77° E) 79°

5. **(2014 AMC 8, #15)**

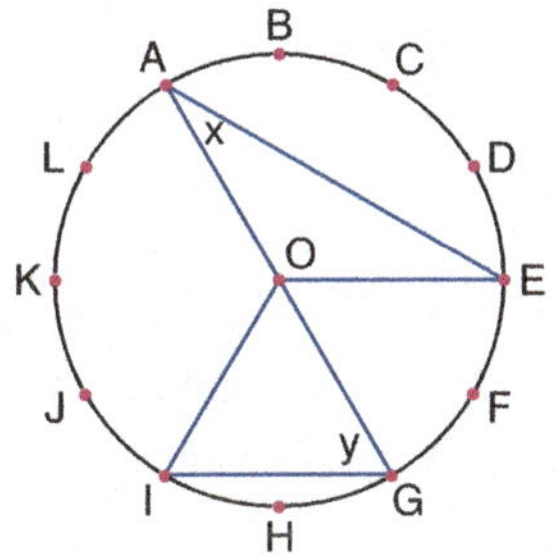

The circumference of the circle with center O is divided into equal arcs, marked the letters A through L as seen below. What is the number of degrees in the sum of the angles x and y ?

A) 75 B) 80 C) 90 D) 120 E) 150

6. **(2007 AMC 10B, #4)**

The point O is the center of the circle circumscribed about DABC, with ∠BOC = 120° and ∠AOB = 140°, as shown. What is the degree measure of ∠ABC?

A) 35 B) 40 C) 45 D) 50 E) 60

7. (MathCON 21 Final GR11, #18)

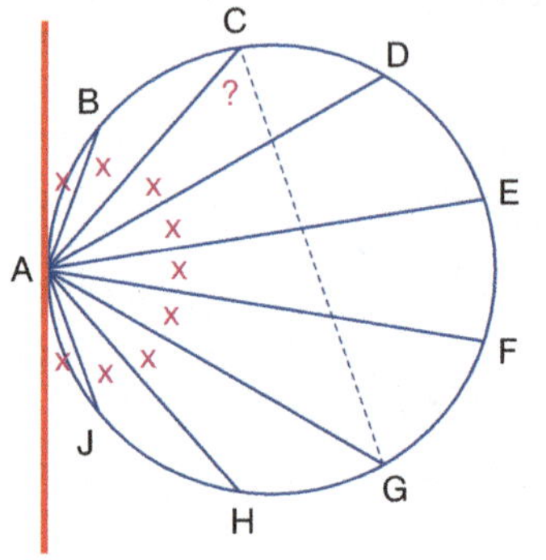

Daniel draws 9 points A, B, C, D, E, F, G, H, and J on a circle and a tangent to the circle at A. If the measurements of all nine angles are equal, then what is the measurement of $\angle ACG$?

(Note that the drawing is not to scale.)

A) 55°   B) 57.5°   C) 60°   D) 62.5°   E) 80°

8. (2015 AMC 12B, #13)

Quadrilateral ABCD is inscribed in a circle with $\angle BAC = 70°$, $\angle ADB = 40°$, $AD = 4$, and $BC = 6$.

What is AC?

A) $3 + \sqrt{5}$ B) 6   C) $\frac{9}{2}\sqrt{5}$ D) $8 - \sqrt{2}$ E) 7

9. (2019 AMC 10A, #13)

Let ΔABC be an isosceles triangle with $BC = AC$ and $\angle ACB = 40°$. Construct the circle with diameter $\overline{BC}$, and let D and E be the other intersection points of the circle with the sides $\overline{AC}$ and $\overline{AB}$, respectively. Let F be the intersection of the diagonals of the quadrilateral BCDE.

What is the degree measure of $\angle BFC$?

A) 90   B) 100   C) 105   D) 110   E) 120

10. (2016 AMC 12B, #23)

Two congruent circles centered at points A and B each pass through the other circle's center. The line containing both A and B is extended to intersect the circles at points C and D. The circles intersect at two points, one of which is E.

What is the degree measure of $\angle CED$?

A) 90   B) 105   C) 120   D) 135   E) 140

11. (MathCON 18 Final GR11, #39)

ABCD is a concave quadrilateral $m\angle ABC = m\angle ADC = 90°$,

$m\angle BAC = 40°$, $m\angle CAD = 20°$, and $|BD| = 6$.

What is the length of $|AC|$?

A) 8   B) $8\sqrt{3}$ C) 6   D) $6\sqrt{3}$ E) $4\sqrt{3}$

12. (MathCON 19 Final GR11, #36)

In a semicircle with center O, $m\angle OAC = 52°$ and $AC = CD = CE$.

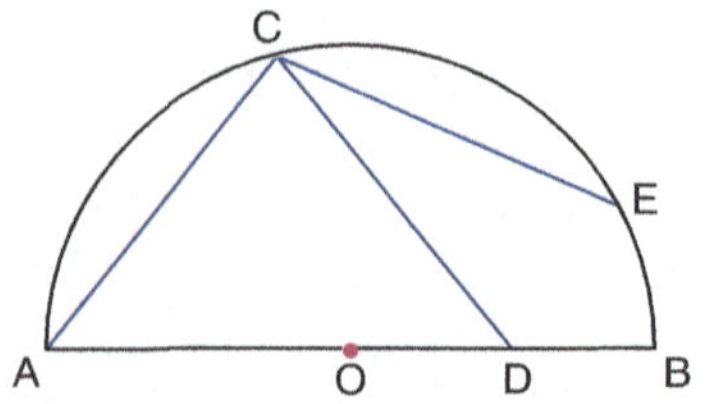

What is $m\angle DCE$?

A) 24°   B) 28°   C) 32°   D) 36°   E) 40°

1.

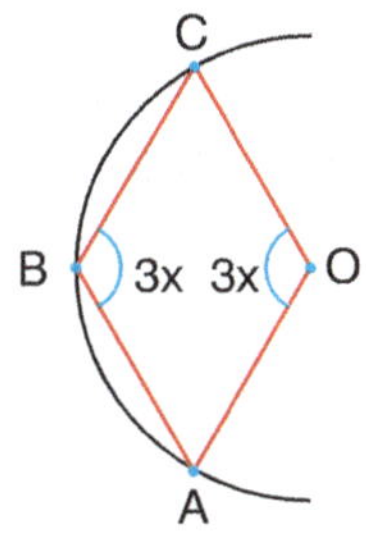

x = ?

2.

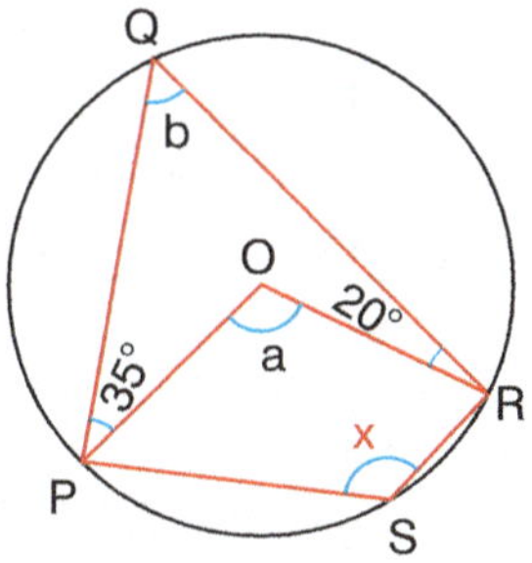

x = ?

3.

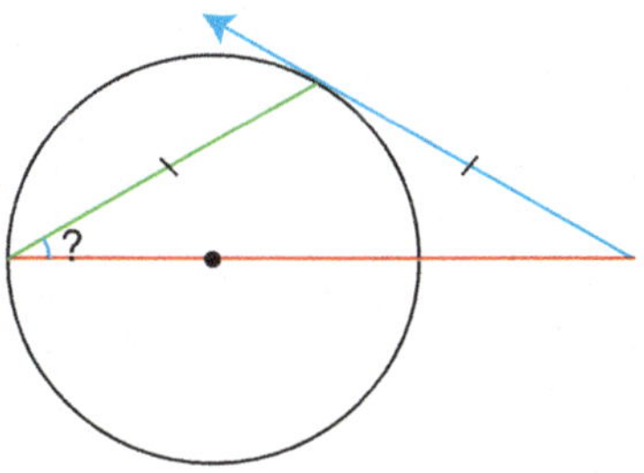

Find the marked angle?

4.

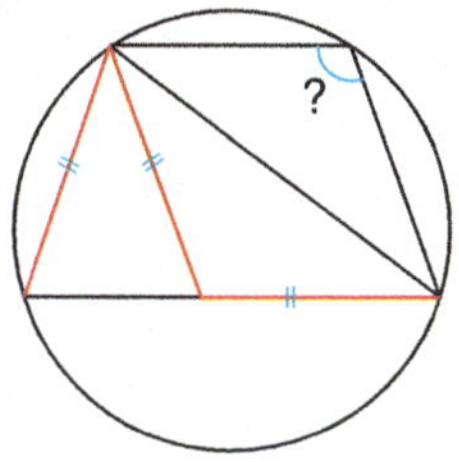

5.

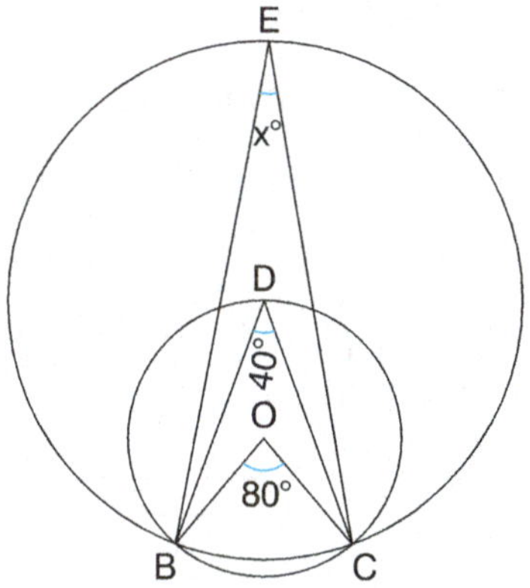

6.

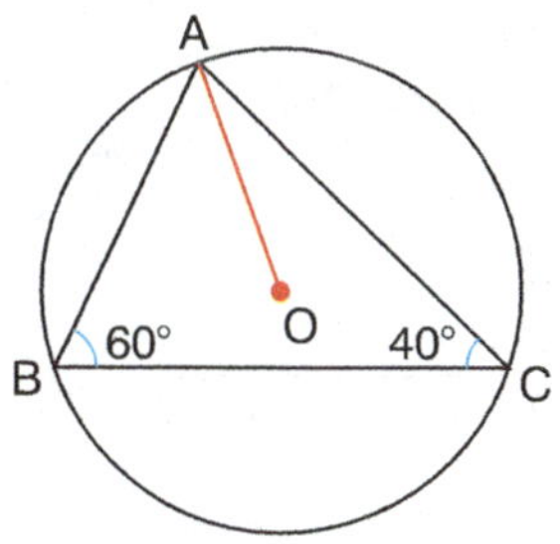

∠BAO?

7.

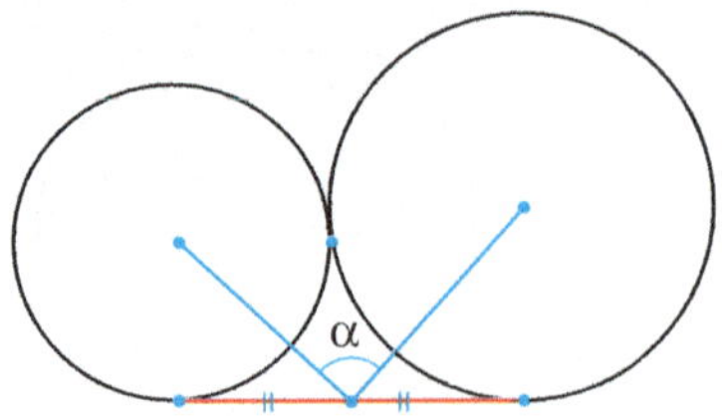

$\alpha = ?$

8. (2014 AMC 8 Problem, 15)

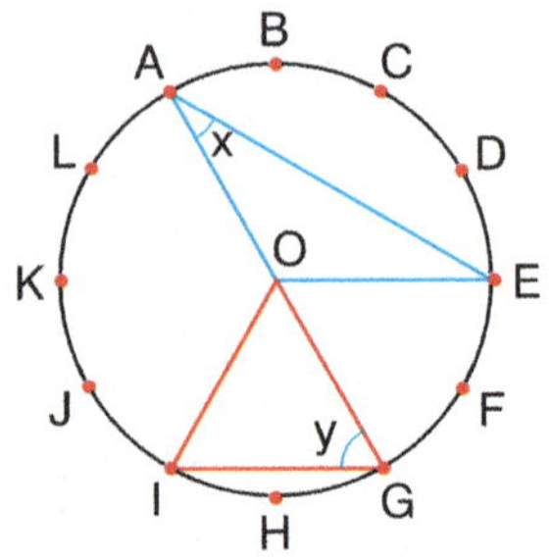

$x + y = ?$

9.

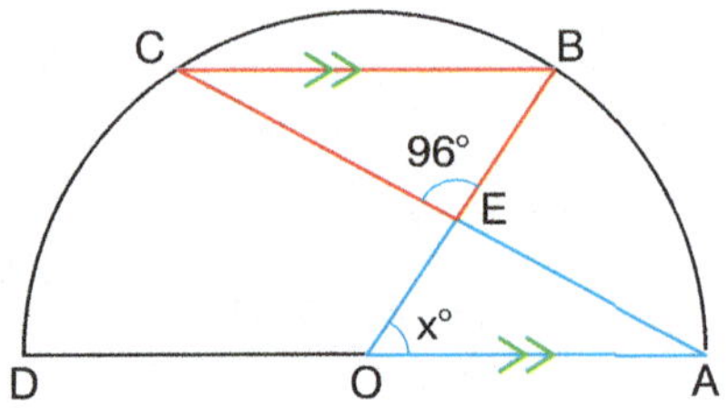

10.

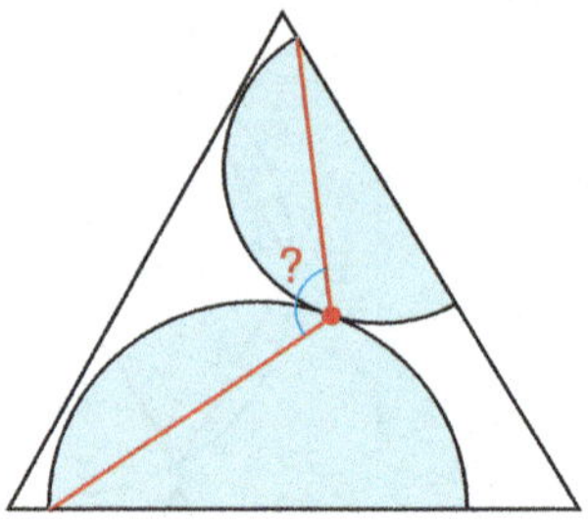

Equilateral triangle.

11.

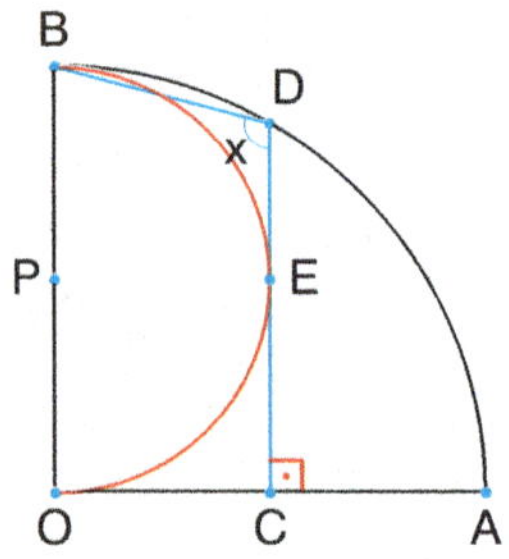

$x = ?$

12.

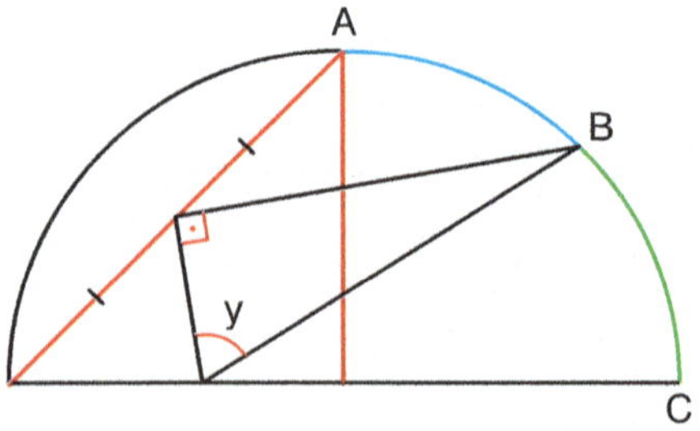

arc AB = arc BC.

Find y?

CHAPTER 2

CIRCLE – LENGTHS

Theorem 1.1

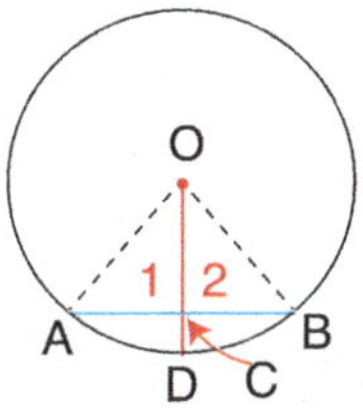

A radius perpendicular to a chord bisects the chord.

Proof

Draw radii $\overline{OA}$ and $\overline{OB}$.

Since all radii of a circle are congruent, OA = OB.

Because $\angle ODA$ and $\angle ODB$ are right angles and OD is common to both triangles, we have $\Delta ODA \cong \Delta ODB$ by the hypotenuse-leg (HL) theorem. Therefore, the corresponding sides are congruent: AD = DB

Hence, $\overline{OD}$ bisects $\overline{AB}$.

Example 1.1

In the figure, $ON \perp KL$ and KL = 10 cm. Find r.

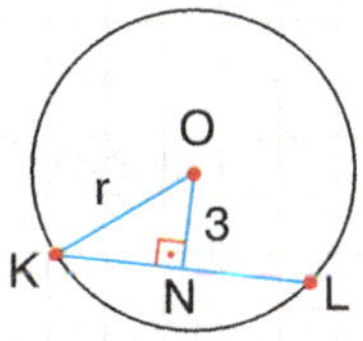

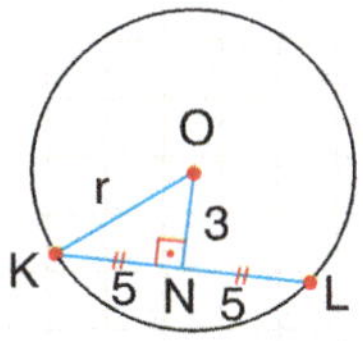

Solution:

Since $ON \perp KL$, by the previous theorem ON bisects KL.

So $KN = NL = \dfrac{KL}{2} = \dfrac{10}{2} = 5$ cm.

By the pythagorean theorem,

$r^2 = 3^2 + 5^2 = 9 + 25 = 34$

$r = \sqrt{34}$ cm.

Example 1.2

In the figure, AB = 10 cm and OH = 4 cm. Find the length OC = x.

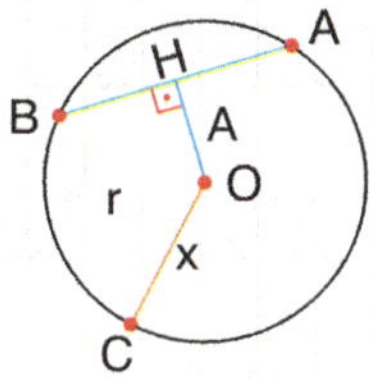

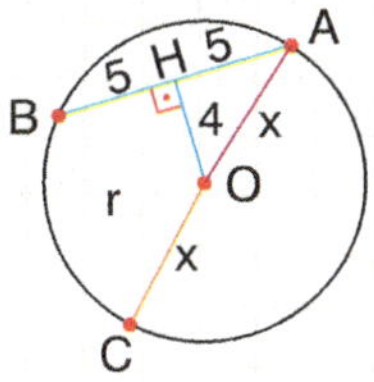

Solution:

Since $OH \perp AB$, it bisects AB.

$AH = \dfrac{AB}{2} = \dfrac{10}{2} = 5$ cm.

By the pythagorean theorem in ΔOHA:

$x^2 = 4^2 + 5^2 = 16 + 25 = 41$

$x = \sqrt{41}$ cm

Property 1.1

If a line is perpendicular to a radius at its endpoint on the circle then the line is tangent to the circle.

Example 1.3

In the figure, PT is tangent to circle and RT = 18. Find the length PT = x.

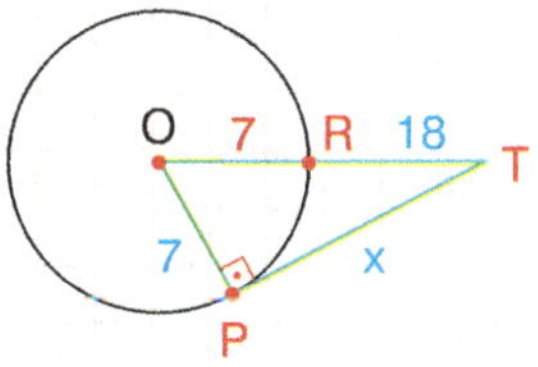

Solution:

Since PT is tangent to circle, $PT \perp OP$.

$7^2 + x^2 = 25^2$

$x^2 = 625 - 49$

$x^2 = 576$

$x = 24.$

Property 1.2

The line connecting the centers of two tangent circles contains the point of tangency and is perpendicular to the common tangent.

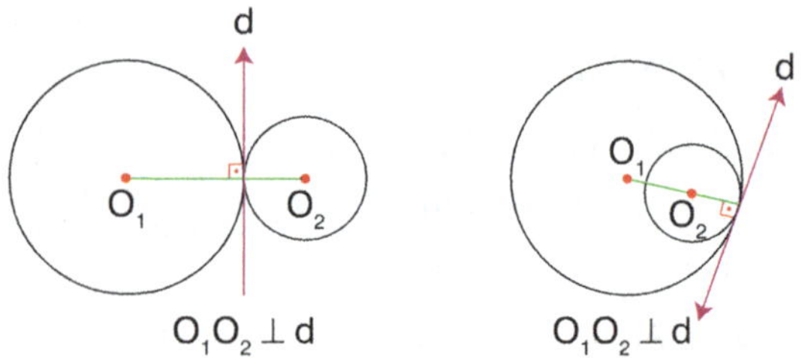

Example 1.4

In the figure, ABCD is a rectangle with AD = 18 cm and AB = 25 cm.

Given that $\odot O_1$ and $\odot O_2$ are externally tangent, find the length r_2.

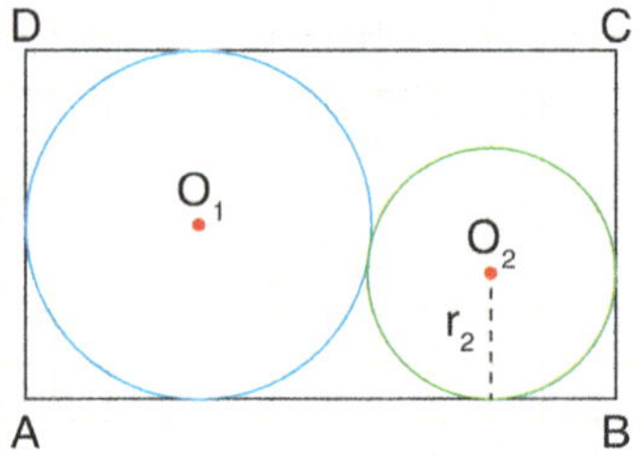

Solution:

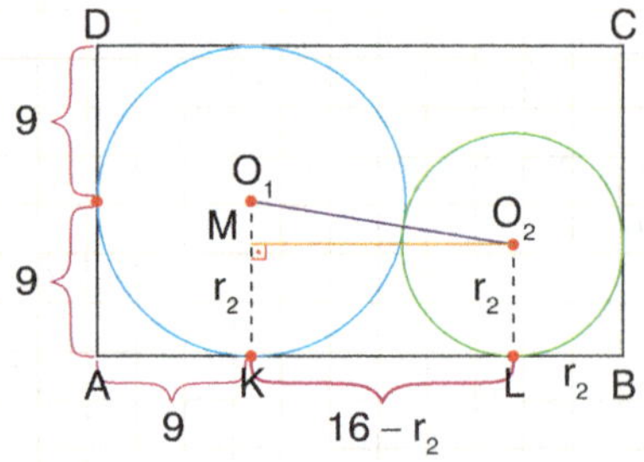

In rectangle KLO_2M: $O_1M^2 + O_2M^2 = (O_1O_2)^2$

Substitute: $(9 - r_2)^2 + (16 - r_2)^2 = (9 + r_2)^2$

Simplify: $r_2^2 - 68r_2 + 256 = 0 \Rightarrow (r_2 - 64)(r_2 - 4) = 0$

Thus, $r_2 = 64$ or 4.

Since AD = 18, $r_2 = 4$ cm.

Theorem 1.2

If a tangent and a secant intersect outside a circle, the product of the external and total secant lengths equals the square of the tangent length. (external) × (whole) = (tangent)^2

$$PA^2 = PC \cdot PB$$

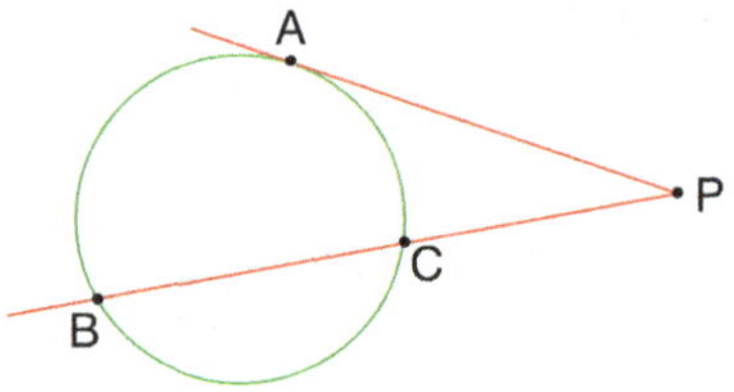

Proof

Both ∠CAP and ∠ABC intercept the same arc, so they are equal. ∠P is common to triangles ΔACP and ΔBAP.

Hence, the triangles are similar (AAA).

By similarity: $\frac{PA}{PB} = \frac{PC}{PA} \Rightarrow PA^2 = PC \cdot PB$

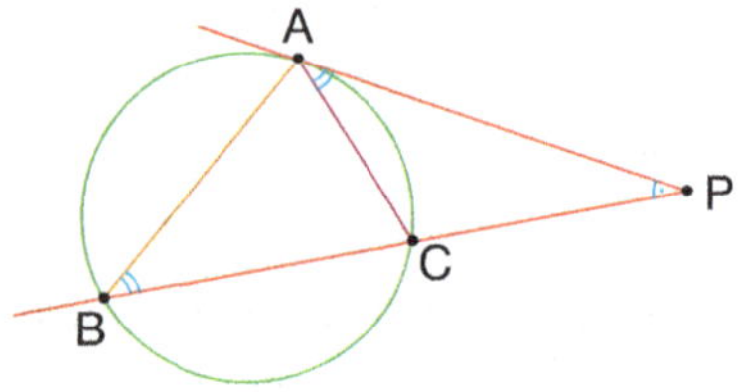

MathTopia Press

Theorem 1.3

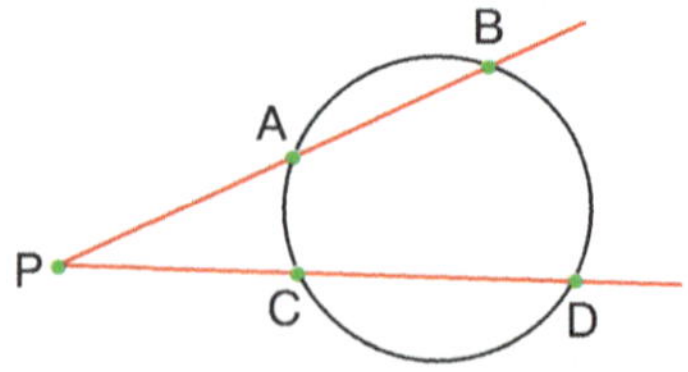

$$PA \cdot PB = PC \cdot PD$$

Theorem 1.4

If two chords intersect inside a circle, the products of their segment lengths are equal. $PA \cdot PC = PB \cdot PD$

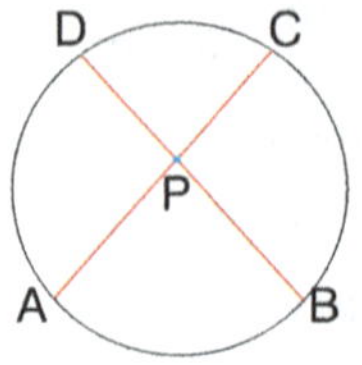

Proof

In the figure, $\angle A$ and $\angle D$ subtend the same arc $\overset{\frown}{BC}$, so the are equal. Similarly, $\angle B$ and $\angle C$ are equal.

Since $\angle APB = \angle DPC$ (vertical angles), triangles ΔAPB and ΔDPC are similar (AAA).

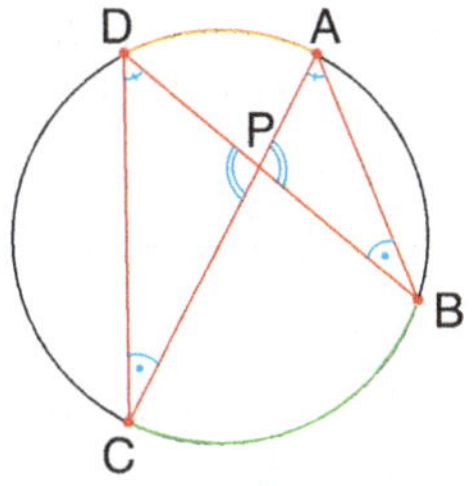

Thus, $\dfrac{PA}{PD} = \dfrac{PB}{PC} \Rightarrow PA \cdot PC = PB \cdot PD.$

Example 1.5

Find the value of $x + y$ in the figure.

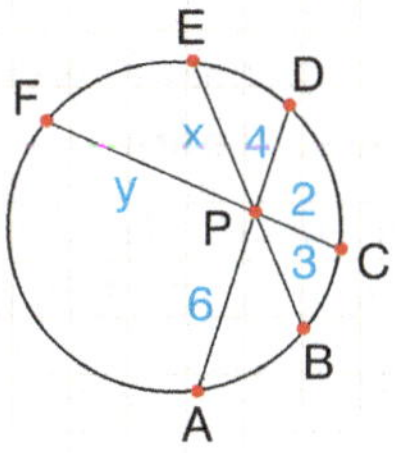

Solution:

By the corollary we have just seen, $PA \cdot PD = PB \cdot PE = PC \cdot PF$, i.e.

$6 \cdot 4 = 3 \cdot x = 2 \cdot y.$

So $x = 8$ cm and $y = 12$ cm,

and $x + y = 8 + 12 = 20$ cm

Example 1.6

In the figure below, RS is the diameter of the circle. The tangent from point P touches the circle at Q.

If PR = 6 cm and PQ = 12 cm, find the radius of the circle.

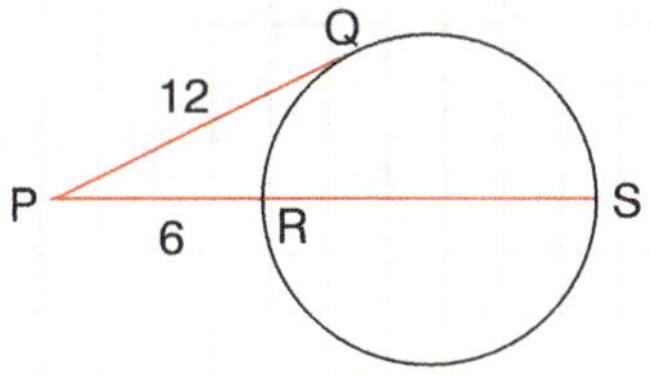

Solution:

By the tangent-secant theorem: $PQ^2 = PR \times PS$

Let the radius be r. Since RS is a diameter, $RS = 2r$.

Then PS ? $PR + RS = 6 + 2r$.

Substitute: $122 = 6(6 + 2r) \rightarrow 144 = 36 + 12r$

$108 = 12r \rightarrow r = 9$

Example 1.7

Two circles are tangent at C. From an external point P:

- The tangent to the circles at C is $PC = x$.
- A secant passes through the small circle with external part 3 and internal part y.

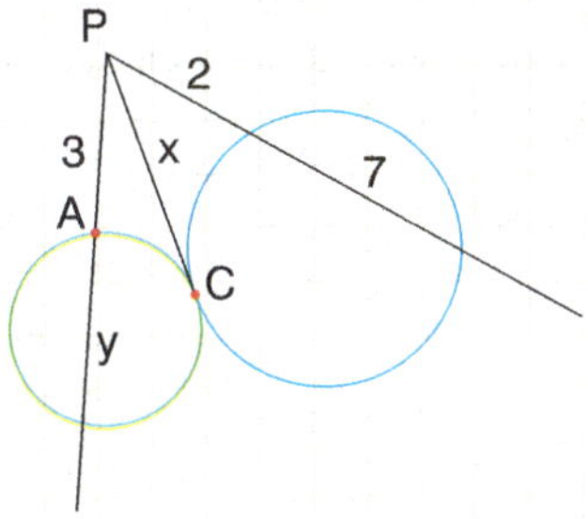

- Another secant passes through the large circle with external part 2 and internal part 7.

Find x and y.

Solution:

For any circle from the same external point P:

$(\text{tangent})^2 = (\text{external}) \times (\text{whole secant})$.

Large circle: $x^2 = 2(2 + 7) = 18 \rightarrow x = 3\sqrt{2}$.

Small circle: $x^2 = 3(3 + y)$. With $x^2 = 18$:

$18 = 3(3 + y) \rightarrow 6 = 3 + y \rightarrow y = 3$.

$x = 3\sqrt{2}$, $y = 3$.

Example 1.8

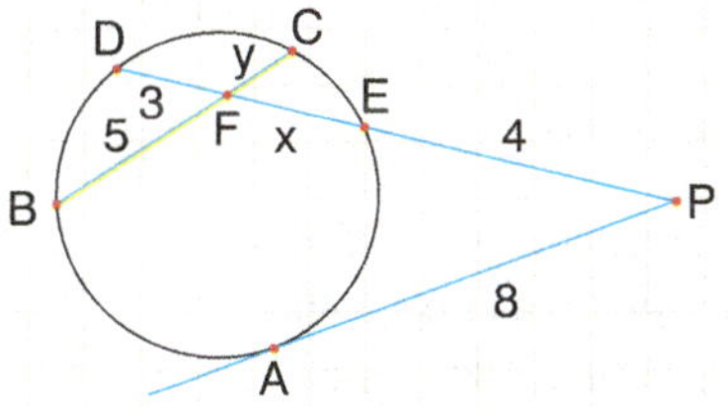

From an external point P, PA = 8 cm is drawn to a circle. A secant P – E – F enters the circle: PE = 4, EF = x, and FD = 3. Find x and y.

Solution:

Intersecting chords at F: chord ED has parts FE = x and FD = 3; the other chord has parts 5 and y.

$(FE)(FD) = (5)(y) \rightarrow x \cdot 3 = 5y$ and $8^2 = 4(7 + x) \Rightarrow x = 9$.

$y = \dfrac{3x}{5} = \dfrac{27}{5}$.

Example 1.9

Given that AB is a common external tangent segment of two externally tangent circles with radii r_1 and r_2, prove that $AB = 2\sqrt{r_1 \cdot r_2}$.

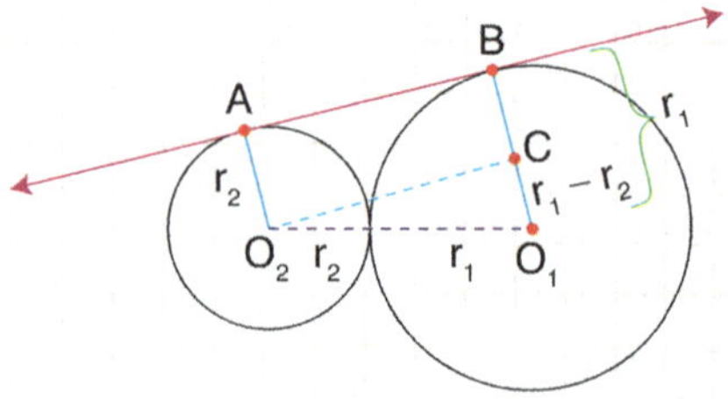

Solution:

$$AB = \sqrt{d^2 - (r_1 - r_2)^2} = \sqrt{(r_1 - r_2)^2 - (r_1 - r_2)^2}$$

$$= \sqrt{\cancel{r_1^2} + \cancel{r_1^2} + 2r_1 r_2 - \cancel{r_1^2} - \cancel{r_2^2} 2r_1 r_2} = \sqrt{4r_1 r_2} = 2\sqrt{r_1 r_2}.$$

Definition 1.1

Orthogonal Circles

Two intersecting circles are called orthogonal circle if they intersect each other at right angles. In orthogonal circles, $r_1^2 + r_2^2 = d^2$, where d is the distance between the two centers.

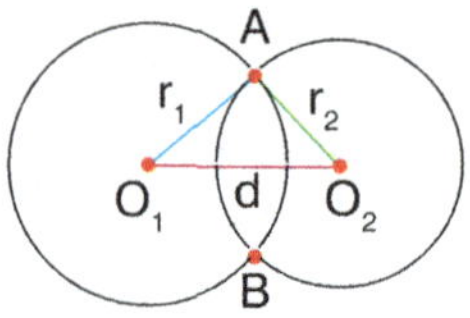

orthogonal circles:
$r_1 \perp r_2$ and $d^2 = r_1^2 + r_2^2$

Example 1.10

A circle with a radius of 8 cm is externally tangent to a circle with a radius of 18 cm. Find the length of a common external tangent.

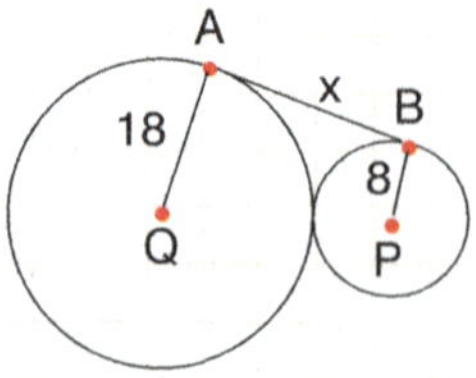

There is a standard procedure for solving a problem involving a common tangent (either internal or external).

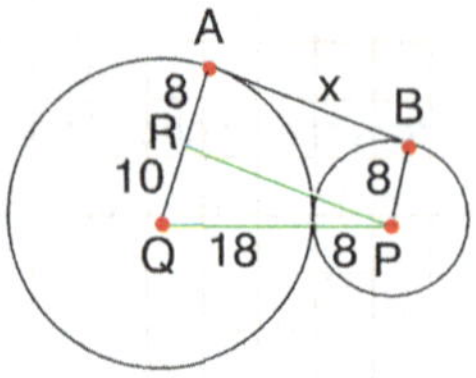

Solution:

In ΔRPQ, $(QR)^2 + (RP)^2 = (PQ)^2$

$10^2 + (RP)^2 = 26^2$, $RP = \pm 24$

Thus, AB = 24 cm.

Example 1.11

Two circles are externally tangent at point T. The first circle has center A and radius 8 cm; the second has center B and radius 2 cm. A common tangent touches the circles at points C and D, respectively. Find the distance CD.

Solution:

Centers A and B are 10 cm apart, with radii 8 and 2. The line through C and D is tangent to both circles, forming right triangles with height difference 8 – 2 = 6.

$CD = \sqrt{10^2 - 6^2} = \sqrt{64} = 8$. CD = 8 cm

Theorem 1.5

Tangential (circumscribed) quadrilateral

If a circle is tangent to all four sides of ABCD, then

AB + CD = BC + AD.

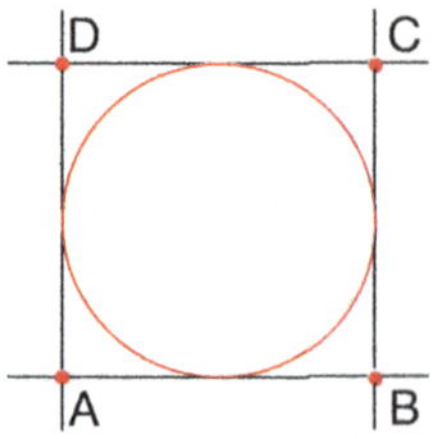

Proof

Let the tangency points on AB, BC, CD, DA be E, F, G, H. From the tangent property,

AE = AH, BE = BF, CF = CG, DG = DH.

Add the first and third: AE + CF = AH + CG. Add the second and fourth: BE + DG = BF + DH.

Then (AB) + (CD) = (AE + BE) (CG + DG) = (AH + BF) + (CF + DH) = (AD) + (BC).

Property 1.3

From a fixed point outside a circle, the two tangent segments to the circle have equal length.

If a circle is tangent to lines PA and PB at A and B, then

PA = PB.

MathTopia Press

Example 1.12

In ΔABC, the incircle with center O is tangent to AC, BC, and AB at E, D, and F, respectively. The segments satisfy AE = 3, EC = x, and BD = y. If the perimeter of ΔABC is 18, find x + y.

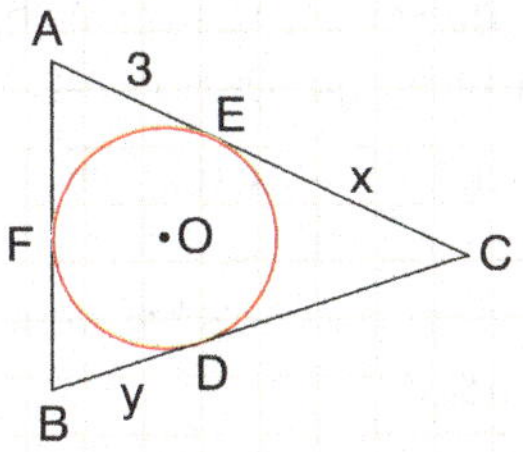

Solution:

For tangents from a common vertex to a circle, the lengths are equal: AF = AE = 3, BF = BD = y, CE = CD = x.

Hence AB = AF + FB = 3 + y, AC = AE + EC = 3 + x, BC = BD + DC = y + x.

The perimeter is $(3 + y) + (3 + x) + (x + y) = 6 + 2x + 2y = 18 \Rightarrow 2x + 2y = 12 \Rightarrow x + y = 6$.

Example 1.13

$|AT| = 16$ cm, $|BT| = 9$ cm

$|OT| = x = ?$ cm

Two tangent circles touch at T; find x.

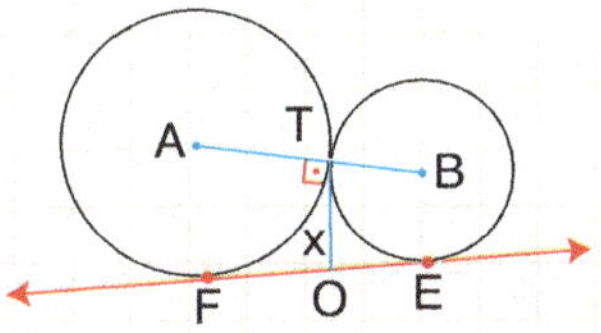

Solution:

$AB^2 = d^2 + (R - r)^2 = (R + r)^2 \Rightarrow d^2 = 4Rr \Rightarrow d = 2\sqrt{Rr}$.

With R = 16, r = 9: $d = 2\sqrt{144} = 24$. Hence EF = 24.

OF = OT = OE = 12

Property 1.3

Angle Bisector Theorem:

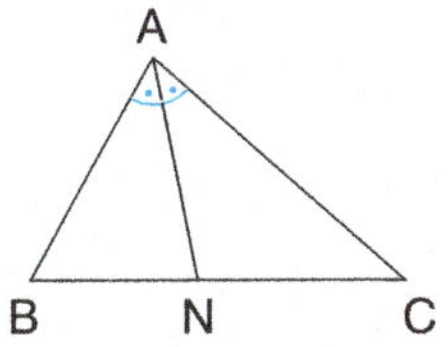

- If AN is the angle bisector then $\frac{AB}{AC} = \frac{BN}{NC}$.
- $AN^2 = AB \times AC - BN \times NC$

Example 1.14

The inscribed circle in the figure is tangent to ΔABC at D, F and G. AB = 10 cm, AC = 14 cm and BC = 12 cm are given. Find DE.

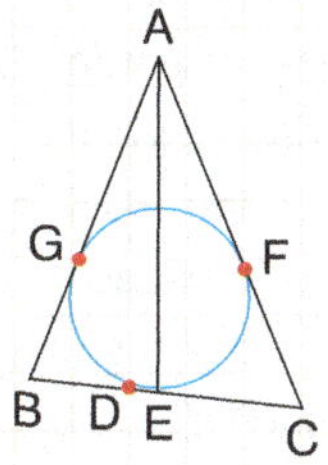

Solution:

Since AE passes through the center of the circle, it bisects $\angle BAC$. By the angle bisector theorem, $\frac{AB}{AC} = \frac{BE}{EC}$, $\frac{10}{14} = \frac{BE}{12 - BE}$, BE = 5 cm.
If 2u = 10 + 14 + 12 then u = 18, and so BG = BD = u – 14 = 18 – 14 = 4 cm.

Therefore DE = 5 – 4 = 1 cm.

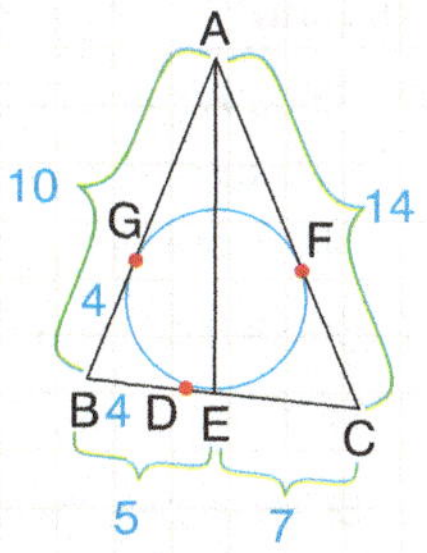

Practice 1

In the figure, the quarter circle centered at O is tangent to ⊙N at A, B and E. AN = 5 cm is given. Find DO.

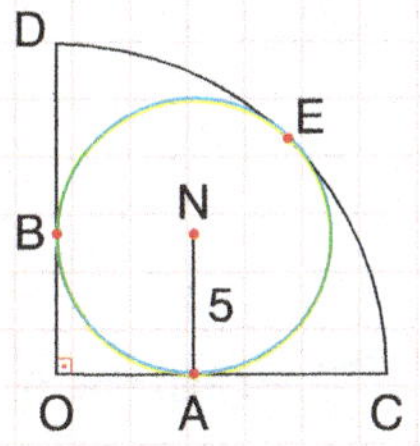

Solution:

Look at the figure. BN and NE are radii, so BN = NE = 5 cm. In ΔOAN, $ON = 5\sqrt{2}$ by the pythagorean theorem. By the previous property, O, N and E are collinear, and $OE = (5\sqrt{2} + 5)$ cm. Since OE and DO are radii of the quarter circle centered at O, $DO = (5\sqrt{2} + 5)$ cm.

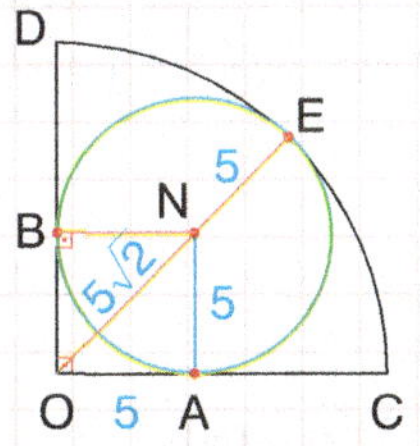

Practice 2

In the figure, ABCD circumscribes ⊙O, AD = 12 cm and CD = 9 cm. Find BE.

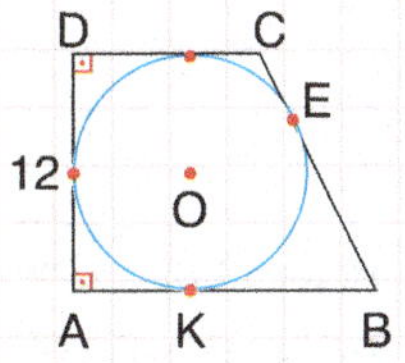

Solution:

Let us construct FG through ⊙O so $FG \perp CD$ and $FG \perp AB$. Then $FG = AD = 2r = 12 \Rightarrow r = 6$ cm. So GC = CD – 6 = 3 cm, and so CE = 3 cm. Now let BE = x. In the right triangle ΔBCK, BC = x + 3, BK = x – 3 and CK = 12 cm. Therefore $(x + 3)^2 = (x - 3)^2 + 12^2$, i.e. x = BE = 12 cm.

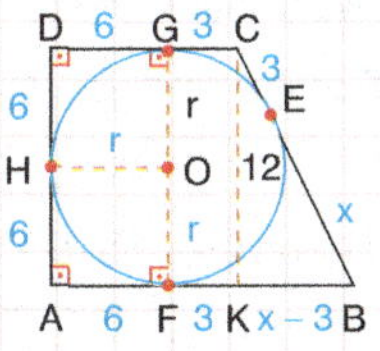

Practice 3

A circle has center O. Points A, B lie on the circle with AB a diameter. A point C lies outside the circle such that AC = AB and BC = 10. The line AC meets the circle again at E. The line through C tangent to the circle touches it at D. Find DE.

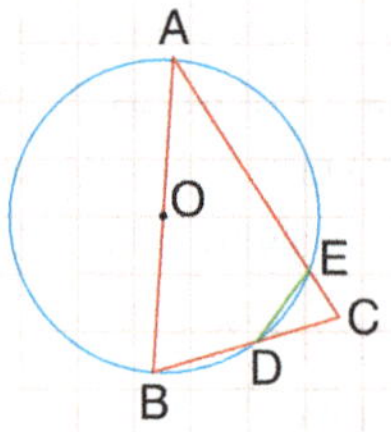

Solution:

1. Since AB = AC, line AD (the bisector from the vertex A) also bisects the arc BD and the arc DE. Therefore, arc BD = arc DE.
2. Equal arcs correspond to equal chords, so BD = DE.
3. From the isosceles ΔABC, the perpendicular from A to BC also bisects BC: $BD = DC = \frac{BC}{2} = 5$.
4. Hence, by step (2), DE = BD = 5.

Practice 4

Adapted from a geometric puzzle originally shared by Catriona Agg (@Cshearer41) on X

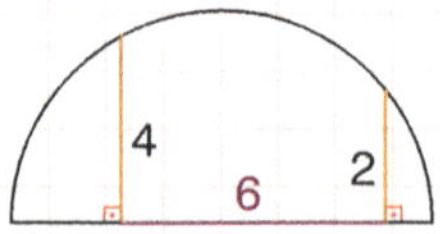

Find radius?

Solution:

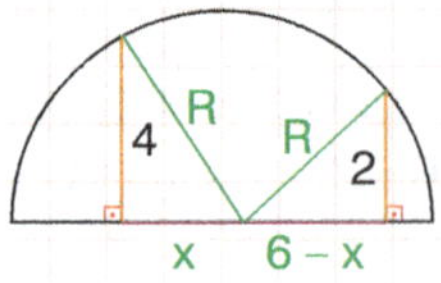

Place O between A and B. Let OA = x; then OB = 6 – x.

Right triangles OA – (height 4) and OB –(height 2) have hypotenuse R: $R^2 = 4^2 + x^2$, $R^2 = 2^2 + (6 - x)^2$.

Equate and solve: $16 + x^2 = 4 + (6 - x)^2 = 4 + 36 - 12x + x^2 \Rightarrow 12 = 36 - 12x \Rightarrow x = 2$.

Then $R^2 = 16 + 2^2 = 20 \Rightarrow R = 2\sqrt{5}$ (≈4.472).

MathTopia Press

Practice 5

Adapted from Senior Mathematical Challenge (UKMT, 1998).

The smaller circle has radius 10 cm. AB is a diameter of this circle. The larger circle has centre A and radius 12 cm.

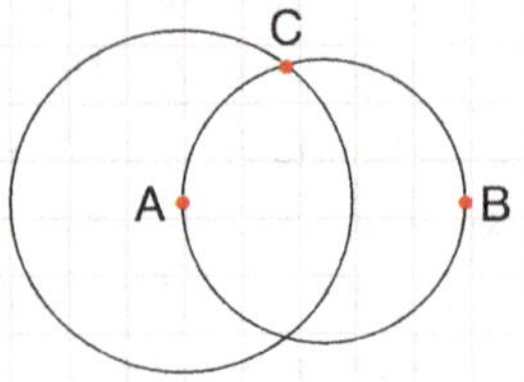

The two circles meet at the point C.

What is the length of the chord CB?

Work Space

Practice 6

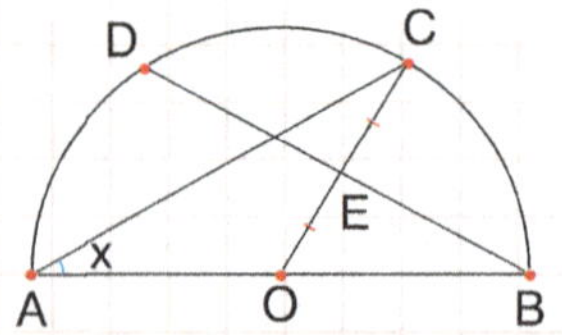

In the figure, AB is the diameter of a semicircle with center O. Point C lies on this semicircle. Points D, E, and B are collinear. The lengths satisfy DE = EB and OE = EC. Also, ∠OAC = x.

Work Space

1.

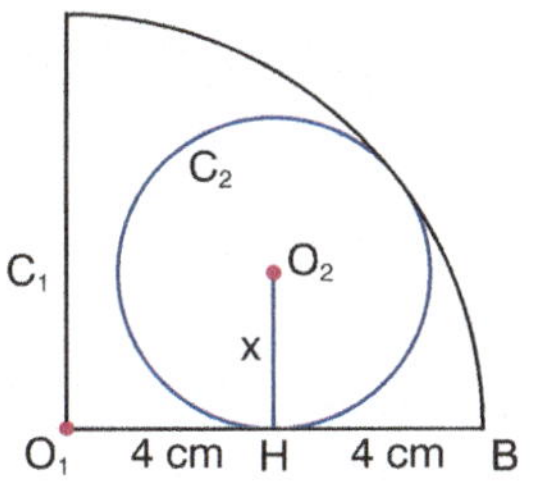

O_1 is a center of C_1 a quarter Circle and O_2 is a center of Circle C_2.

Given that $O_1H = HB = 4$, find $O_2H = x$.

2. A diameter of a circle measures 26 cm and the lengths of two chords parallel to this diameter are 10 and 24cm. What is the shortest distance between these two chords?

3.

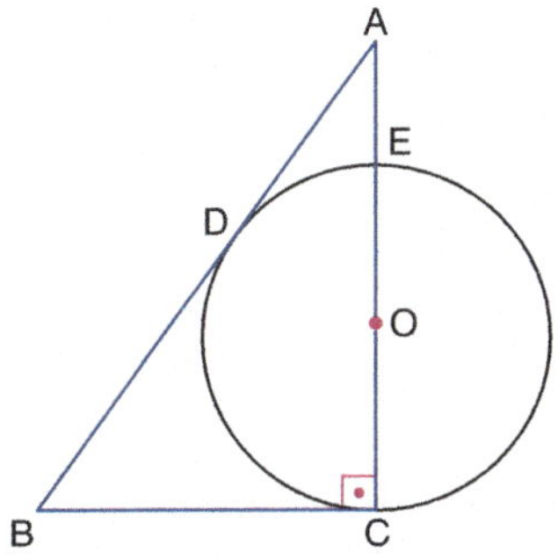

In the figure, circle O tangent to quadrilateral EBCD.

Given that $AE = 1$ and $AD = 3$ find $BC = x$.

4.

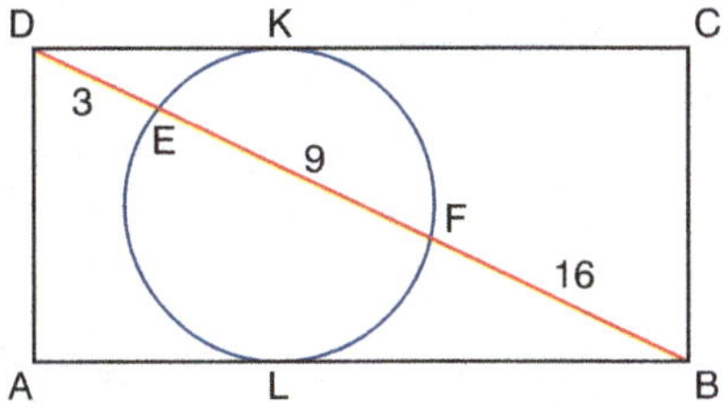

A circle is inscribed in a rectangle as shown.

Given that $DE = 3$, $EF = 9$, and $FB = 16$ find DC.

5.

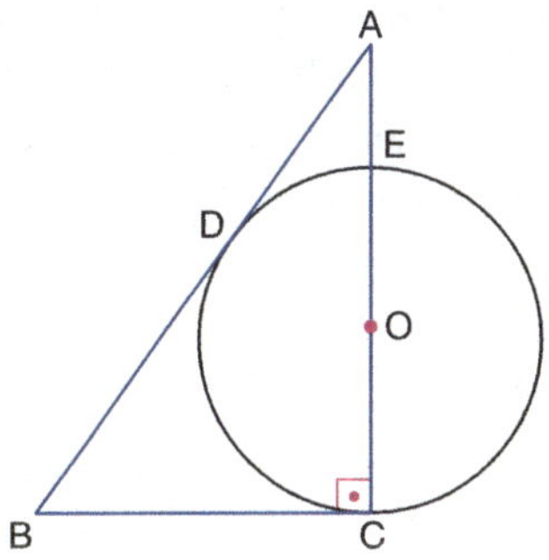

In the figure points D and C are tangent to the circle with diameter of CE.

If $3EC = 2BC$ then what is $\frac{AE}{AD}$?

6. A circle O is tangent to line BC at point F. OB and OC intersect the circle at points E and D, respectively. E and D are not between OB and OC.

Given that $BO \perp OC$, $BF = 8$, and $FC = 6$, find BE.

7.

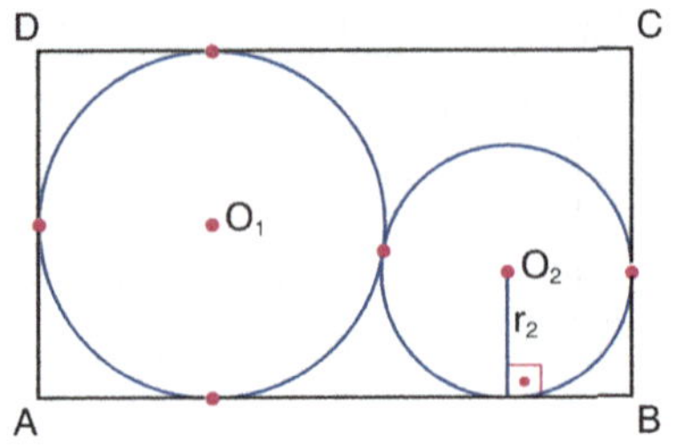

In the figure, ABCD is a rectangle with AD = 18 and AB = 25.

Given that two circles are externally tangent, find the radius of the small circle(r_2).

8. The track shown is uniformly wide, has wto straight sections, rwo semicircular ends, and dimesions as shown. How many yards longer is the track's outer perimeter than its inner perimeter?

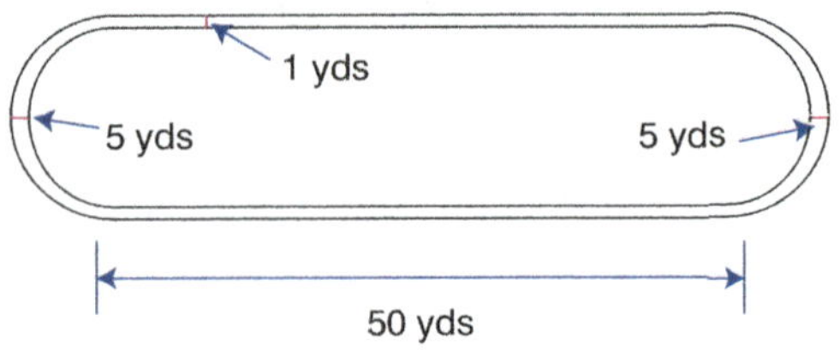

9.

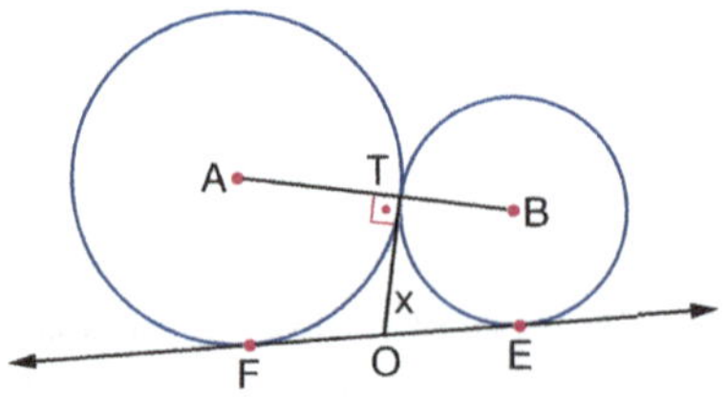

Two cricles are externally tangent at T.

Given that AT = 16 and BT = 9, find OT = x.

10.

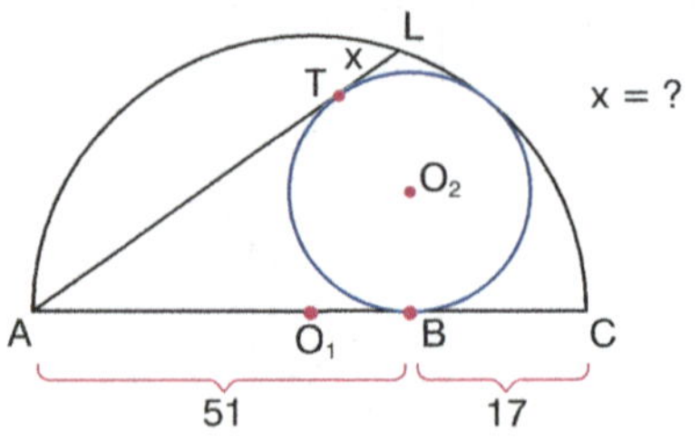

In the figure, T and B are tangent points of circle O_2.

In circle O_1, AB = 51 and BC = 17.

What is TL = x?

11.

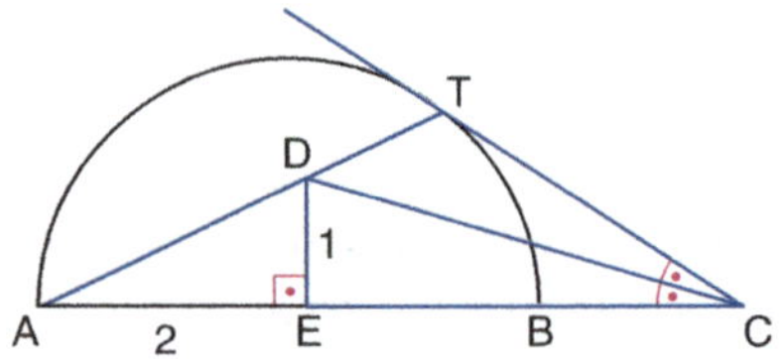

In the figure ray CT is tangent to circle at T and E is a point of tangency.

Given that DE⊥AC, DE = 1, and AE = 2, What is EC?

12.

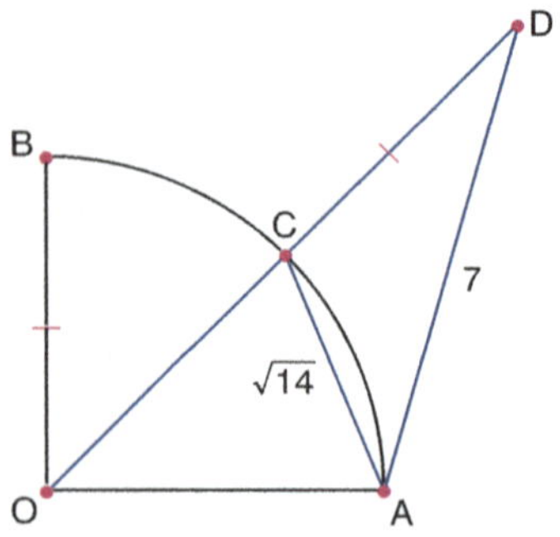

A quarter circle's radius is equal to CD as shown.

Given that AC = $\sqrt{14}$ and AD = 7, find the radius of the circle.

1.

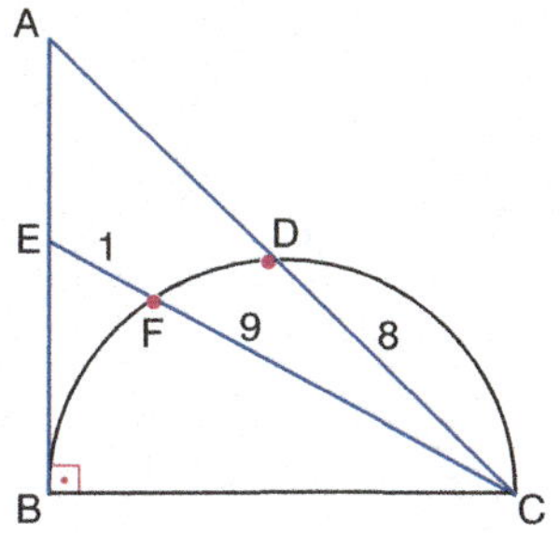

ABC is a right triangle with $\angle B = 90°$ BC is a diameter of the circle. Given that EF = 1, FC = 9, and DC = 8.

What is AD?

2.

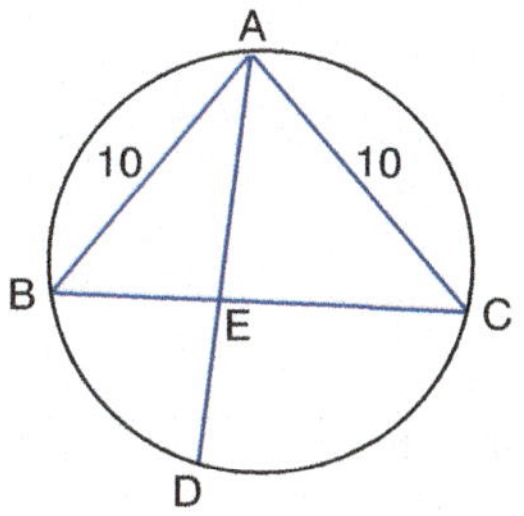

$AD \cap BC = \{E\}$. ABC is an isosceles triangle with

AB = AC = 10. Given that $ED = \frac{9}{2}$.

What is AE?

3.

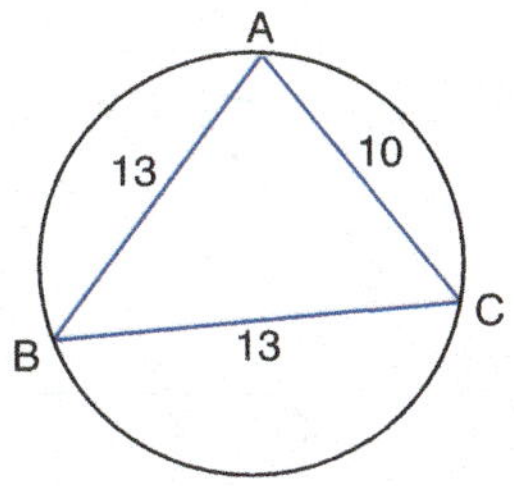

In the figure, ABC is an isosceles triangle with

AB = BC = 13 and AC = 10.

Find the radius of the circle.

4.

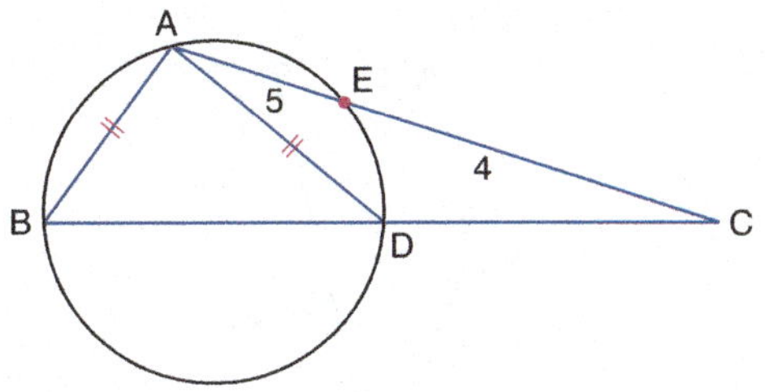

If AE = 5, EC = 4, and AB = AD.

Find AB.

5.

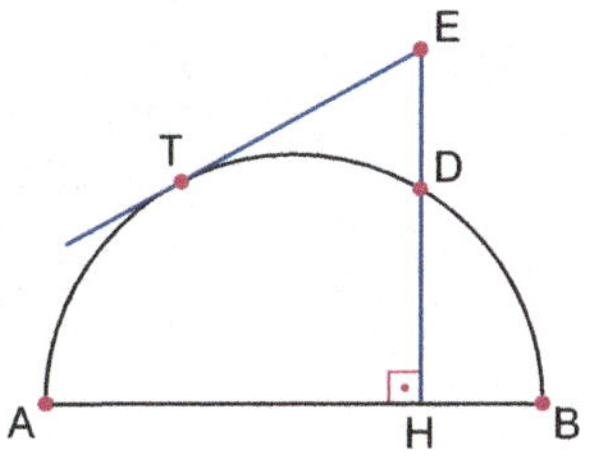

AB is a diameter, T is a tangent of the circle.

Given that ED = 1, DH = 4. Find ET.

6.

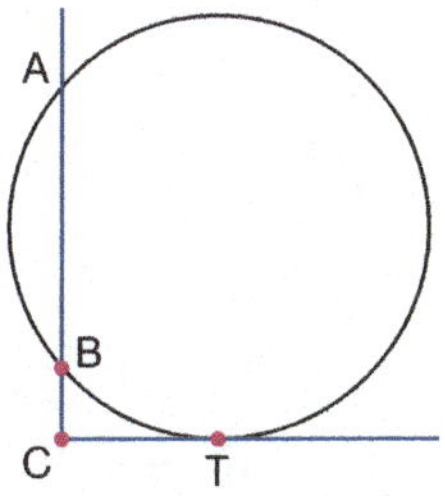

T is a tangent point. $AC \perp CT$.

Given that CT = 6 and AB = 5.

What is the the radius of the circle?

7.

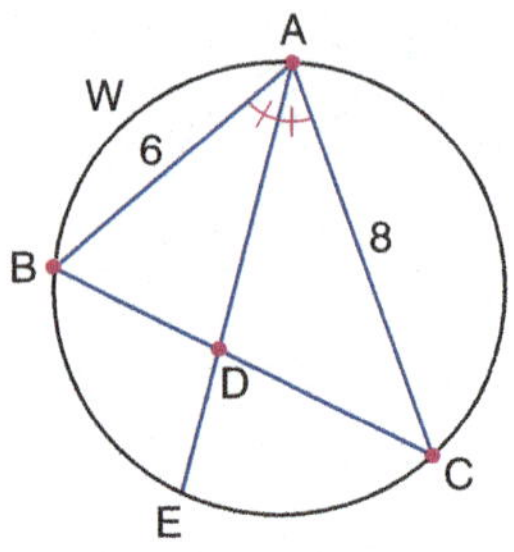

A, B, and C on a circle w and ∠BAE = ∠EAC .

Given that AB = 6, AC = 8, and AD = 4.

What is AE?

8.

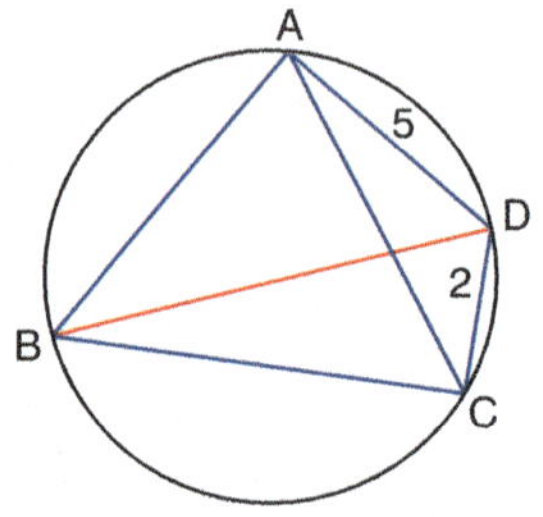

ABC is a equilateral triangle AD = 5 and DC = 2.

What is BD?

9.

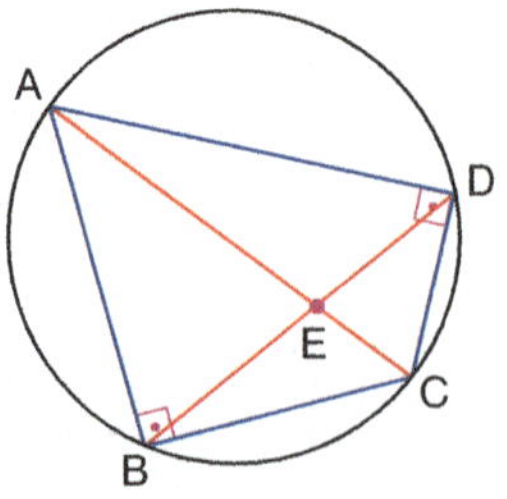

In the figure, AD⊥DC, AB⊥BC, and ∠BAD = 30°.

Given that BE = x, DE = y, and AC = z.

What is z in terms of x and y?

10.

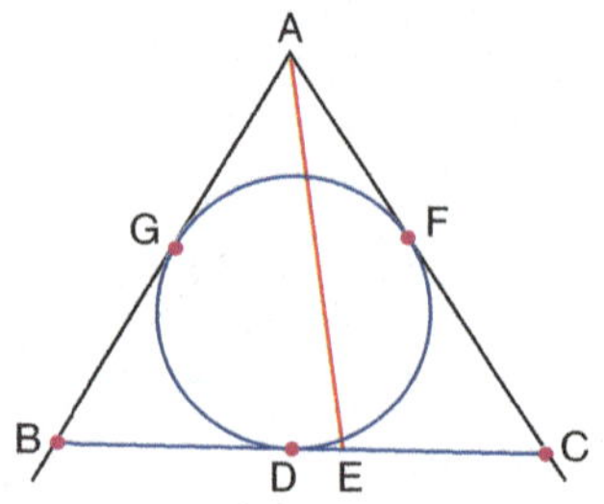

The incribed circle is tangent to Δ ABC at D, F, and G and AE is angle bisector.

AB = 10, AC = 14, and BC = 12.

Find DE.

11.

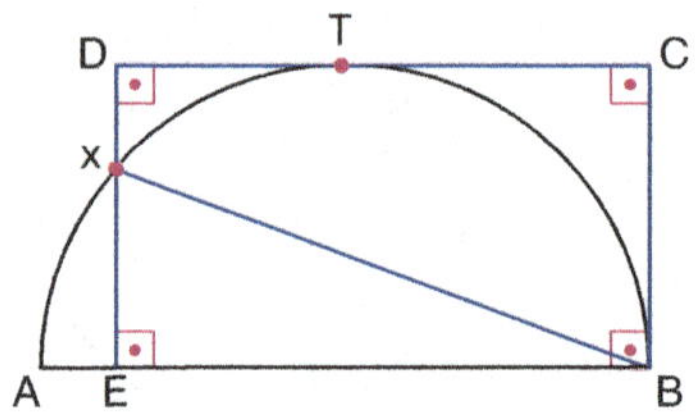

In the figure, AB is a diameter and EDCB is a rectangle. with T is a point of tangency. If XB = 6 then what is the area of the rectangle?

12.

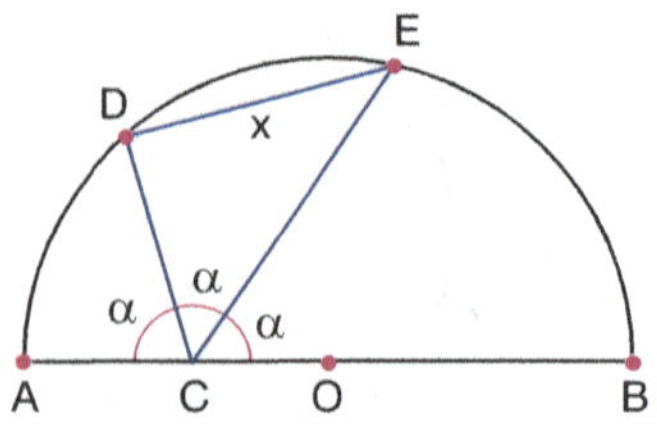

In the figure, AB is a diameter and

m∠ACD = m∠DCE = m∠BCE.

Given that AC = 4 and CB = 8 find DE.

1. (CWML 09-10 Meet 3, #12)

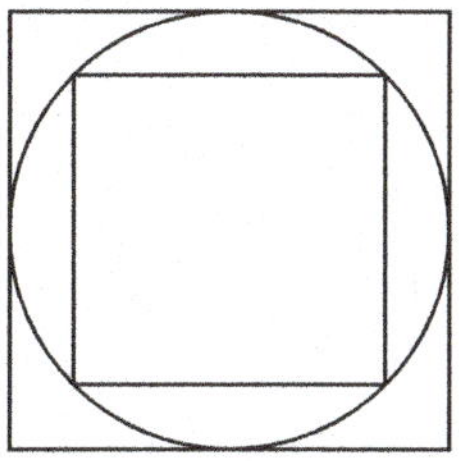

Assume that we are given a square with sides s units long. A circle is inscribed in the square and then another square is inscribed in the circle as shown. Find the length, in terms of s, of a side of the inscribed square. Give the exact answer.

2. (MathCON 19 Final GR12, #12)

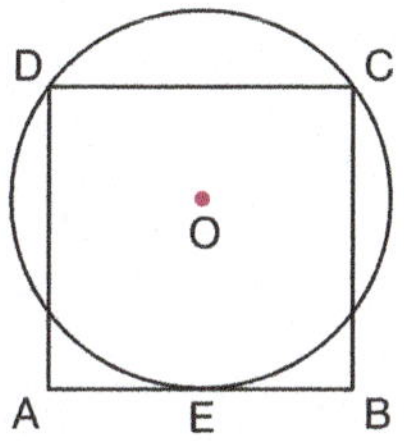

Suppose ABCD is a square with sides of length 2 inches long. If a circle tangent to AB at E contains C and D, find the radius of the circle in inches.

A) 1 B) $\frac{5}{4}$ C) $\frac{5}{2}$

D) $\frac{6}{5}$ E) None of the preceding

3. (CWML 13-14 Meet 3, #10)

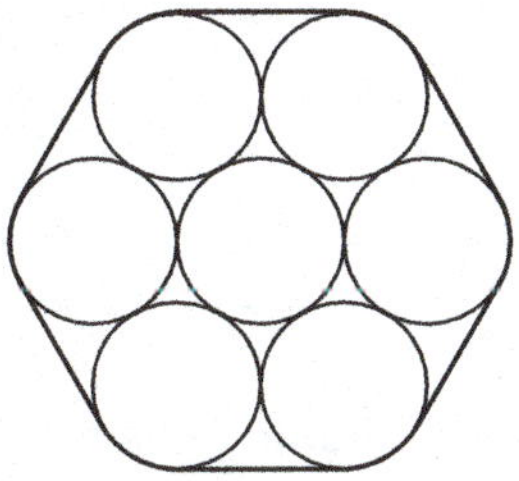

Seven circular discs each with radius 4 inches are tangent to each other and are arranged as shown in the figure. What is the length (in inches) of the shortest cord that could go around them?

4. (2006 AMC 12A, #16)

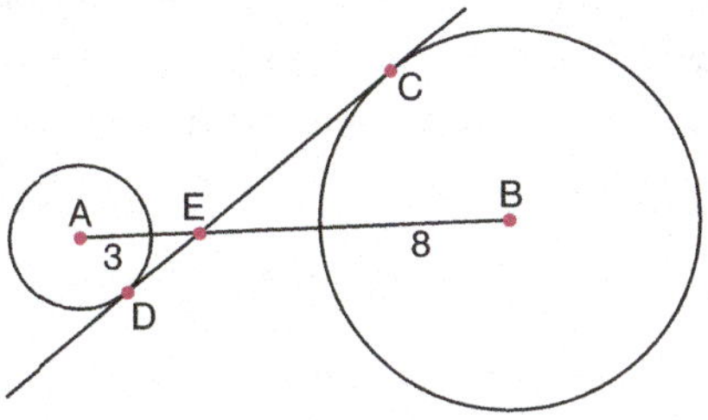

Circles with centers A and B have radii 3 and 8, respectively. A common internal tangent intersects the circles at C and D, respectively.

Lines AB and CD intersect at E, and AE = 5.

What is CD?

5. (2015 AMC 12B, #13)

Rectangle ABCD is inscribed in a semicircle with diameter FE, as shown in the figure.

Let DA = 16, and let FD = AE = 9.

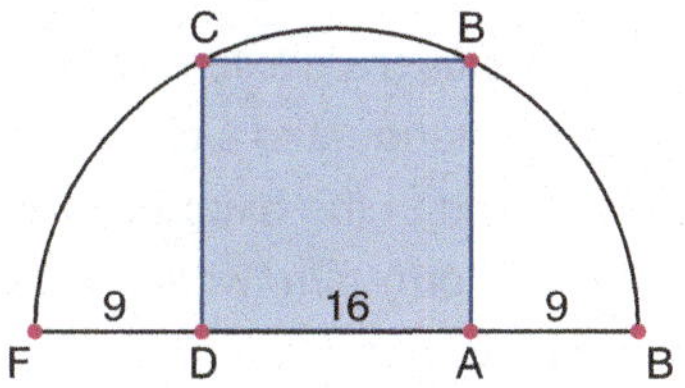

What is area of ABCD?

A) 240 B) 248 C) 256 D) 264 E) 272

6. (2004 AMC 12A, #18)

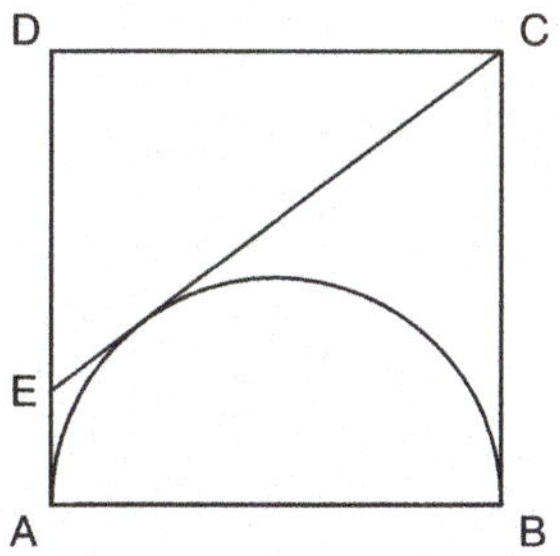

Square ABCD has side length 2. A semicircle with diameter AB is constructed inside the square, and the tangent to the semicircle from C intersects side AD at E.

What is the length of CE?

A) $\frac{(2+\sqrt{5})}{2}$ B) $\sqrt{5}$ C) $\sqrt{6}$ D) $\frac{5}{2}$ E) $5-\sqrt{5}$

7. (2012 AMC 10A, # 11)

Externally tangent circles with centers at points A and B have radii of lengths 5 and 3, respectively. A line externally tangent to both circles intersects ray AB at point C.

What is BC?

A) 4 B) 4.8 C) 10.2 D) 12 E) 14.4

8. (2018 AMC 10A, #15)

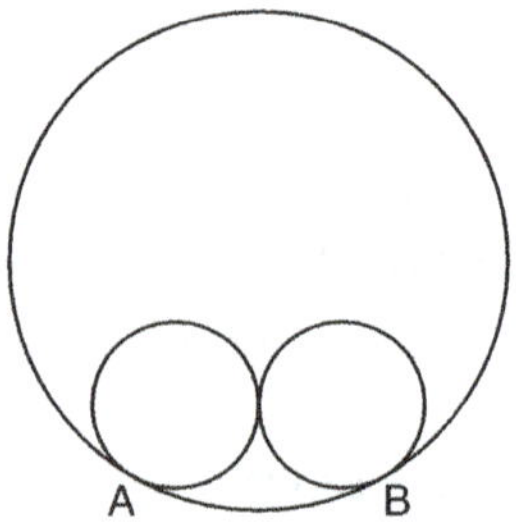

Two circles of radius 5 are externally tangent to each other and are internally tangent to a circle of radius 13 at points A and B, as shown in the diagram. The distance AB can be written in the form m/n, where m and n are relatively prime positive integers.

What is m + n?

A) 21 B) 29 C) 58 D) 69 E) 93

9. (MathCON 21 Final GR11, #22)

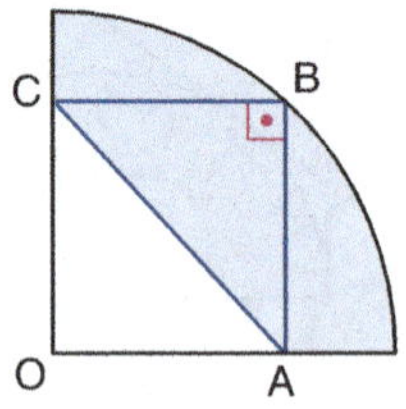

The quarter circle shown has radius 8 cm.
Rectangle OABC has perimeter 16 cm.
If the perimeter of the shaded region is $(4\pi + A)$ cm.

What is the value of A?

A) 8 B) $8\sqrt{2}$ C) $10\sqrt{2}$ D) 12 E) 16

10. (2004 AMC 10A, #23)

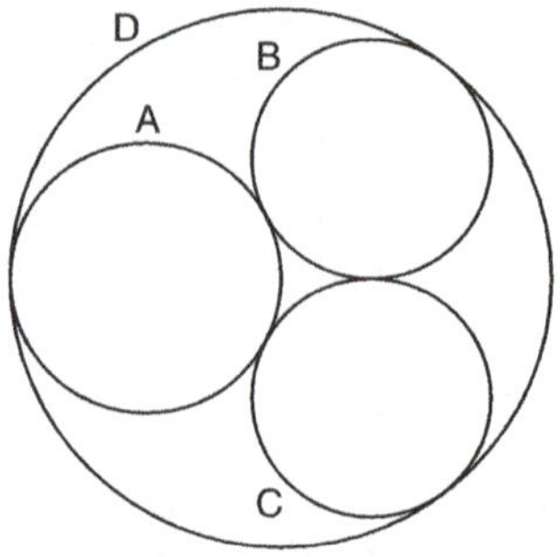

Circles A, B, and C are externally tangent to each other and internally tangent to circle D and circle B and C are congruent. Circle A has radius of 1 and passes through the center of D.

What is the radius of circle B?

A) $\frac{2}{3}$ B) $\frac{\sqrt{3}}{2}$ C) $\frac{7}{8}$ D) $\frac{8}{9}$ E) $\frac{(1+\sqrt{3})}{3}$

11. (2016 AMC 10A, #25)

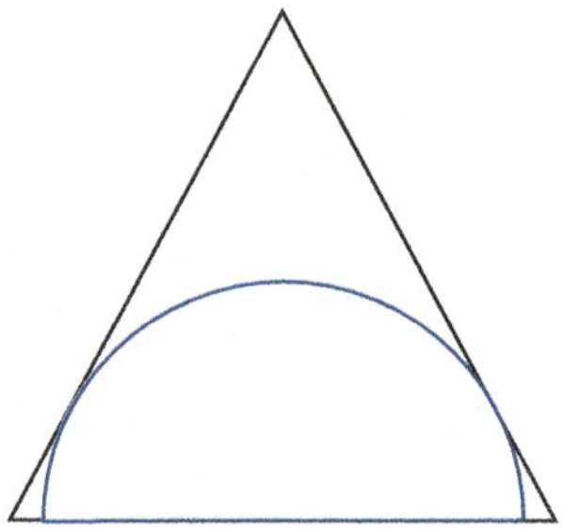

A semicircle is inscribed in an isosceles triangle with base 16 and height 15 so that the diameter of the semicircle is contained in the base of the triangle as shown.

What is the radius of the semicircle?

12. (MathCON 21 Final GR11, #30)

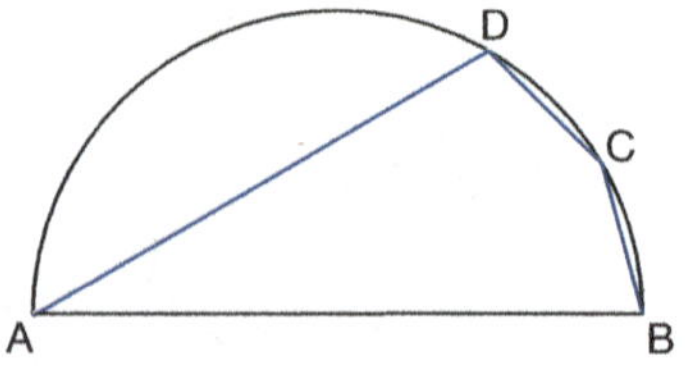

In the semicircle shown, AB is the diameter with AB = 12 cm.

Point C and D are on the semicircle such that BC = CD = 3 cm.

What is the length of AD, in cm?

1.

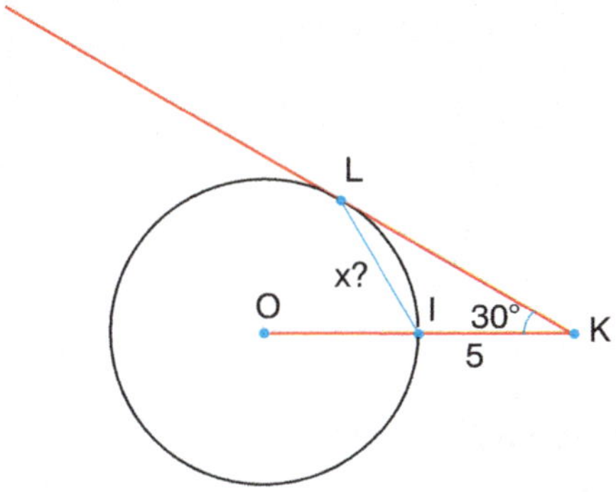

4.

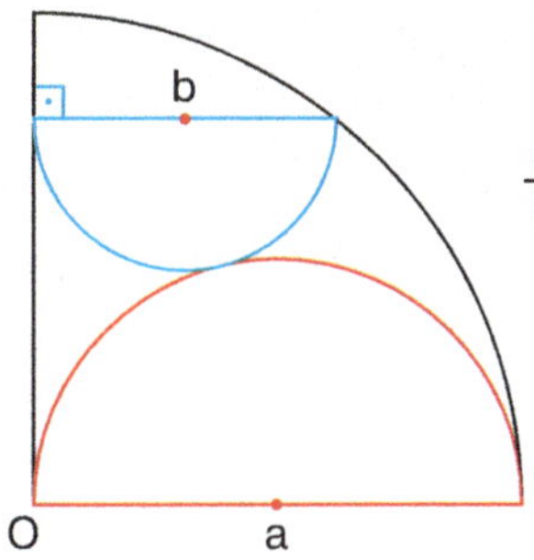

$\frac{a}{b} = ?$

2.

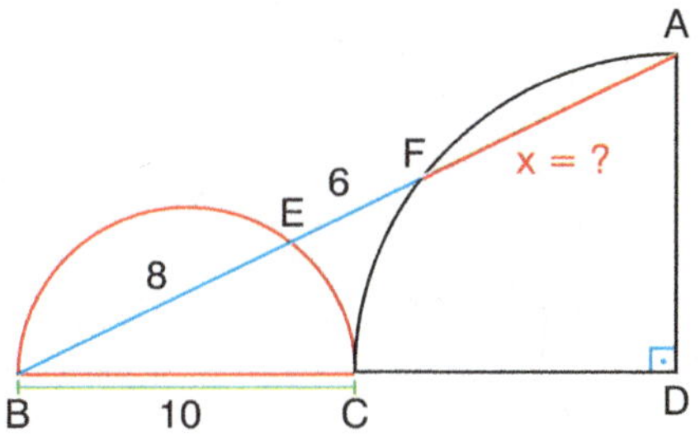

5.

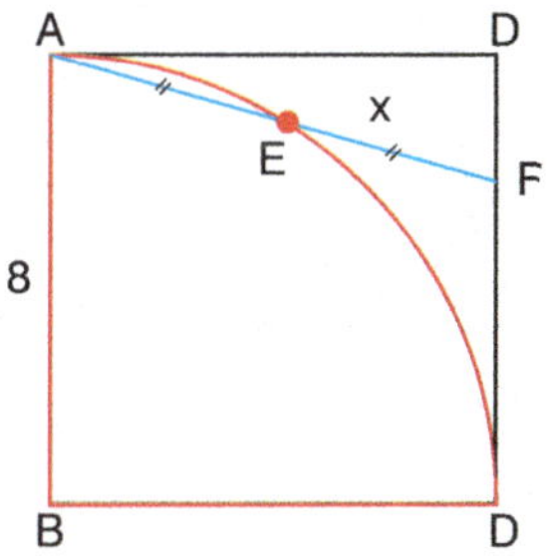

AD = x = ?

MathTopia Press

3.

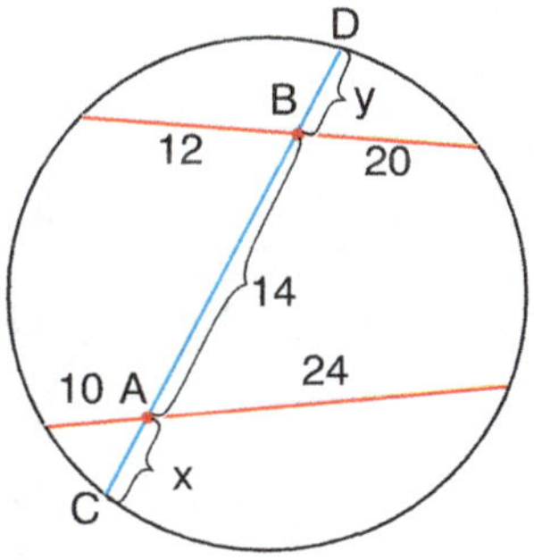

x + y = ?

6.

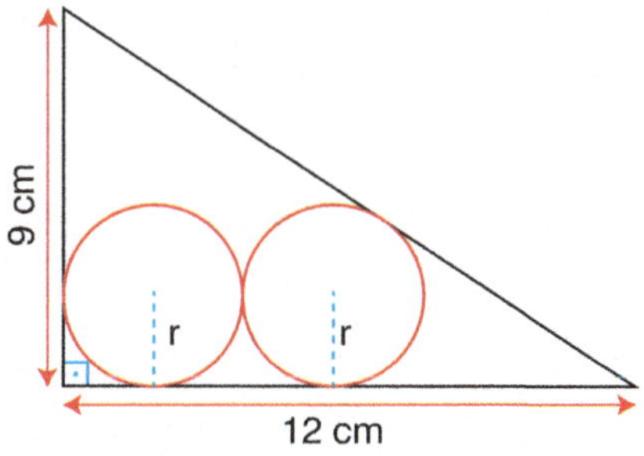

r = ?

7.

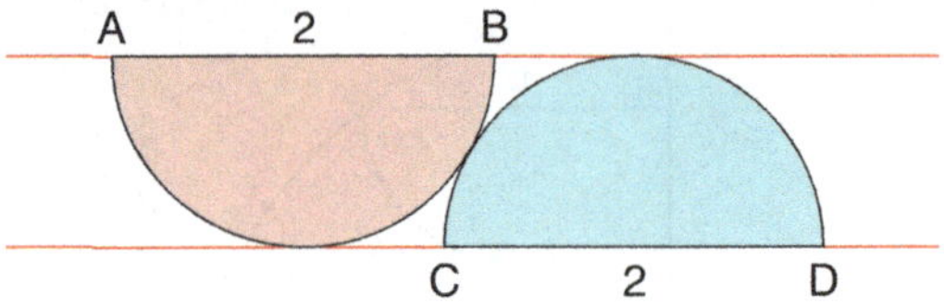

AD = ?

8.

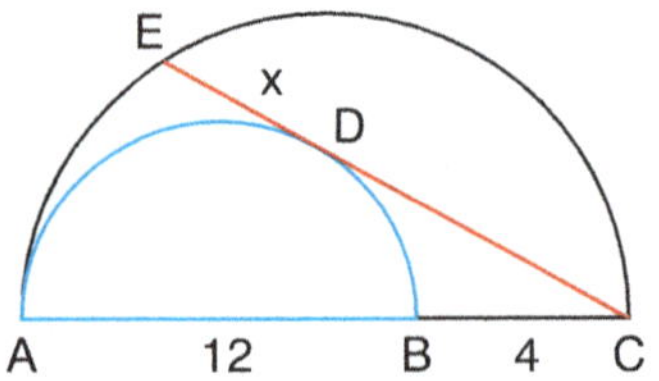

x = ?

9.

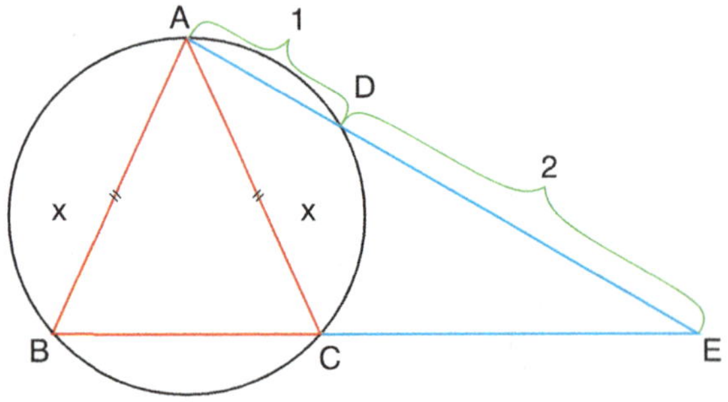

x = ?

10.

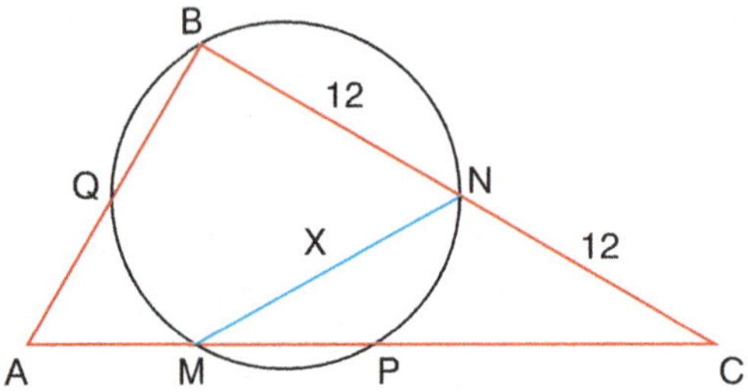

N, P, Q are midpoints.

X = ?

11.

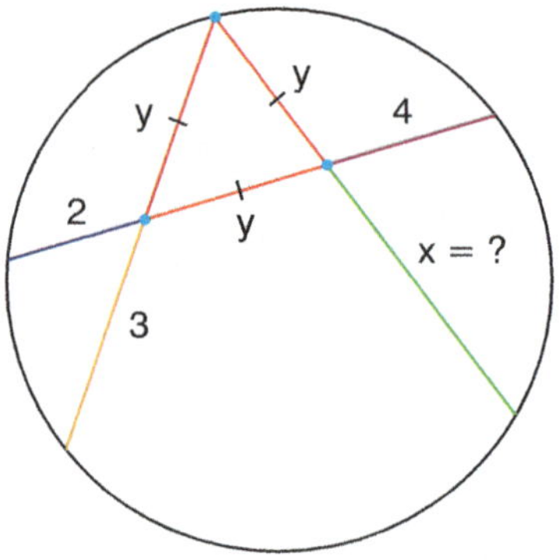

12.

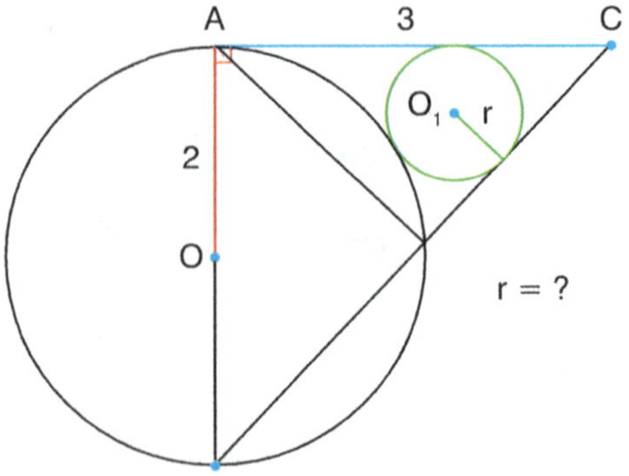

r = ?

CHAPTER 3

CIRCLE AREAS

Definition 1.1

- **Annulus:** The region between two concentric circles with radii $R > r$; its area equals the difference of their areas, $\pi(R^2 - r^2)$.
- **Ring Sector:** The area between two concentric sectors sharing the same central angle.
- **Apothem:** The perpendicular distance from the center of a regular polygon to one of its sides; used when approximating circular area using inscribed polygons.
- **Circumference:** The total boundary length of the circle, $C = 2\pi r$.

Theorem 1.1

Area of a Circle

Statement. The area of a circle with radius r is

$$A_{circle} = \pi r^2.$$

Proof

Proof (dissection "pizza-slices"). Partition the circle into 2n equal sectors; rearrange alternating sectors to form an almost-rectangle of height r and base approximately $\frac{1}{2}C = \pi r$. As $n \Rightarrow \infty$, gaps vanish, giving area $A = \text{base} \times \text{height} = (\pi r) \cdot r = \pi r^2$.

Theorem 1.2

Sector Area is proportional to Angle

Statement. For a central angle θ (in degrees) in a circle of radius r, the sector area is

$$A_{sector} = \frac{\theta}{360^\circ} \pi r^2.$$

Proof

Partition a full circle (360°) into equal sectors; area is proportional to the number of sectors, hence proportional to θ. Scaling from the full area πr^2 yields the formula.

Theorem 1.3

Annulus (Ring) Area

Statement. For concentric circles of radii $R > r$,

$$A_{annulus} = \pi(R^2 - r^2).$$

Proof

Subtract inner circle area from outer circle area:

$$\pi R^2 - \pi r^2 = \pi(R^2 - r^2).$$

Property 1.1

Ring Sector Area

Two concentric sectors with the same central angle θ and radii $R > r$ form a ring sector of area

$$A = \frac{\theta}{360^\circ} \pi (R^2 - r^2).$$

Example 1.1

A circle of radius 10 cm has a central angle 54°. Find the sector area.

Solution:

$A = \frac{54}{360}\pi(10)^2 = \frac{3}{20}\pi \cdot 100 = 15\pi \text{ cm}^2.$

Example 1.2

A circular ring has outer radius 12 cm and inner radius 7 cm. How much area is painted?

Solution:

$A = \pi(12^2 - 7^2) = \pi(144 - 49) = 95\pi \text{ cm}^2.$

Example 1.3

In a circle of radius r = 9 cm, a chord is c = 12 cm. Find the minor segment area.

Solution:

$\frac{c}{2} = 6,\ r^2 - (r - h)^2 = 6^2 \Rightarrow 81 - (9 - h)^2 = 36$

$(9 - h)^2 = 45 \Rightarrow 9 - h = \sqrt{45} = 3\sqrt{5} \Rightarrow h = 9 - 3\sqrt{5}.$

Triangle base-height area along the chord:

$A_\Delta = \frac{1}{2}c(r - h) = \frac{1}{2} \cdot 12(9 - (9 - 3\sqrt{5})) = 6 \cdot 3\sqrt{5} = 18\sqrt{5}.$

Central angle (degrees) from isosceles triangle:

$\cos\frac{\theta}{2} = \frac{r - h}{r} = \frac{3\sqrt{5}}{9} = \frac{\sqrt{5}}{3}$

(use calculator or table if allowed).

$$\frac{\theta}{2} \approx 41.81° \Rightarrow \theta \approx 83.62°$$

Sector area

$$A_{sector} = \frac{83.62}{360}\pi(92) \approx 18.81\pi$$

Minor segment area

$$A_{segment} = A_{sector} - A_\Delta \approx 18.81\pi - 18\sqrt{5}$$

Theorem 1.4

Area Ratios from Similarity of Sectors

Statement. In the same (or congruent) circle(s), sectors with equal central angles are similar and their areas are in the ratio of the squares of radii:

$$\frac{A_1}{A_2} = \left(\frac{r_1}{r_2}\right)^2.$$

Proof

Scale factor $k = \frac{r_1}{r_2}$ multiplies all lengths by k, so areas scale by k^2.

Example 1.4

A square of side 20 cm has quarter-circles of radius 10 cm cut out at two opposite corners (sharing the same center at the corner). Find the remaining area.

Solution:

Start with the square: 400. Two quarter-circles combine to a half-circle $\frac{1}{2}\pi(10)^2 = 50\pi$.

Remaining $= 400 - 50\pi \text{ cm}^2$.

MathTopia Press

My Notes:

Example 1.5

An annulus is the region between two concentric circles. The concentric circles in the figure have radii b and c, with $b > c$. Let $\overline{OX}$ be a radius of the larger circle, let $\overline{XZ}$ be tangent to the smaller circle at Z, and let $\overline{OY}$ be the radius of the larger circle that contains Z. Let $a = XZ$, $d = YZ$, and $e = XY$. What is the area of the annulus?

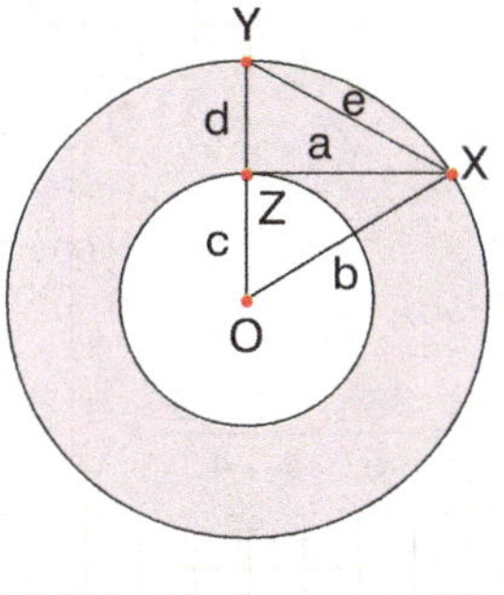

A) πa^2 B) πb^2 C) πc^2 D) πd^2 E) πe^2

Solution:

The area of the larger circle is πb^2 and the area of the smaller circle is πc^2, so the area of the annulus is $\pi(b^2 - c^2)$. By the pythagorean theorem on ΔOXZ, $b^2 - c^2 = a^2$, so the area simplifies to (A)πa^2.

Practice 1

In cyclic kite ABCD with $AB = AD = 10$, $BC = CD = 6$, find the circle's radius R and the area of the circle.

Solution:

Diagonals of a kite are perpendicular; use ptolemy to find $AC \cdot BD = AB.\ CD + AD \cdot BC = 2 \cdot 10 \cdot 6 = 120$.

Also $\text{Area}_{\Delta ABC} = \frac{1}{2}AC \cdot (BD/2)$ and similarly for ΔADC, giving total kite area $= \frac{1}{2}AC \cdot BD = 60$.

For a cyclic quadrilateral with perpendicular diagonals, Area $= \frac{pq}{2} = 60$ where $p = AC$, $q = BD$. With $pq = 120$, consistent.

Now $\text{Area}_{kite} = \frac{abc}{4R}$ can be applied by splitting into two isosceles triangles (or use $K = \frac{1}{2}ab \sin C$ and circumradius relation $a = 2R\sin A$, etc.). A computation yields $R = 8$.

Therefore $A_{circle} = \pi R_2 = 64\pi$.

Example 1.6

The circles with radius 2 are mutually tangent. What is the total area of the circles and the region bounded by them, as shown in the figure?

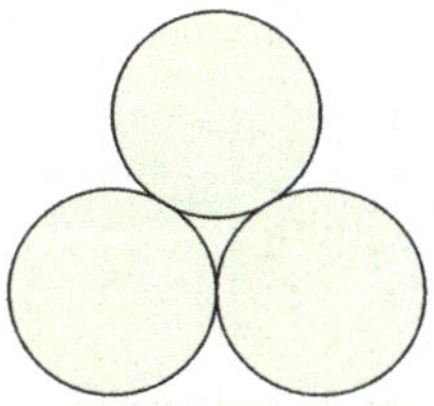

Solution:

We can draw the lines between the radii; this gives us an equilateral triangle with side length 4. We can draw a height of length $2\sqrt{3}$ by 30-60-90 triangles, so the area of the triangle is $\frac{1}{2} \cdot 4 \cdot \left(2\sqrt{3}\right) = 4\sqrt{3}$. Then, we have 300 degree sectors of circles, so the area of each sector is $\frac{300}{360}\pi 2^2 = \frac{10}{3}\pi$. In total, the circular parts add up to $3 \cdot \frac{10}{3}\pi = 10\pi$ so our answer is $10\pi + 4\sqrt{3}$.

Practice 2

Quadrilateral ABCD with side lengths $AB = 7$, $BC = 24$, $CD = 20$, $DA = 15$ is inscribed in a circle. The area interior to the circle but exterior to the quadrilateral can be written in the form $\frac{a\pi - b}{c}$, where a, b, and c are positive integers such that a and c have no common prime factor. What is $a + b + c$?

A) 260 B) 855 C) 1235 D) 1565 E) 1997

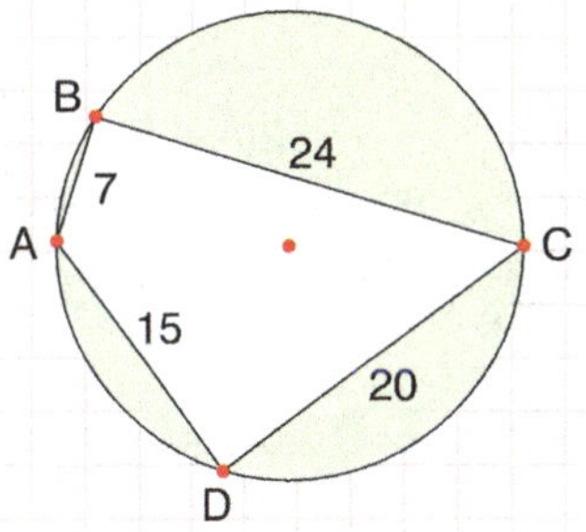

Solution:

Opposite angle of every cyclic quadrilateral are supplementary, so $\angle B + \angle D = 180^\circ$.

We claim that AC = 25. We can prove it by contradiction:

- If AC < 25, then $\angle B$ and $\angle D$ are both acute angles. This arrives at a contradiction.
- If AC > 25, then $\angle B$ and $\angle D$ are both obtuse angles. This arrives at a contradiction.

By the inscribed angle theorem, we conclude that $\overline{AC}$ is the diameter of the circle. So, the radius of the circle is

$r = \frac{AC}{2} = \frac{25}{2}$. The area of the requested region is

$\pi r^2 - \frac{1}{2} \cdot AB \cdot BC - \frac{1}{2} \cdot AD \cdot DC = \frac{625\pi}{4} - \frac{168}{2} - \frac{300}{2} = \frac{625\pi - 936}{4}$.

Therefore, the answer is a + b + c = (D) 1565.

Practice 3

Adapted from a geometric puzzle originally shared by Catriona Agg (@Cshearer41) on X

The center of the semicircle lies on the perimeter of the quarter circle. What's the shaded area?

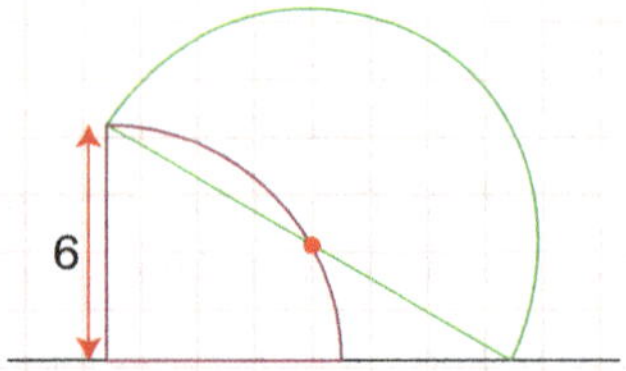

Solution - 1:

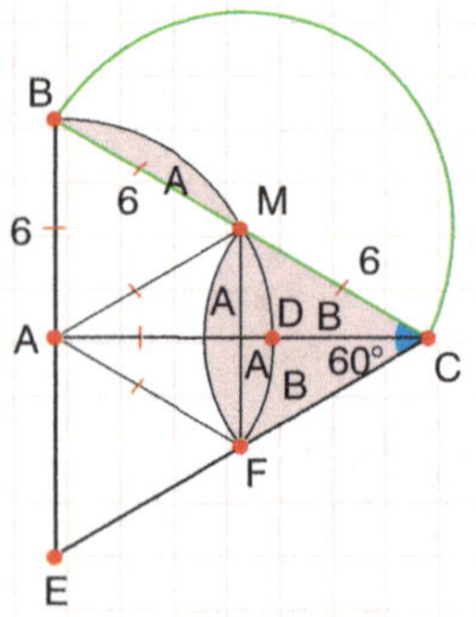

$$2(A + B) = \frac{\pi \cdot 6^2}{6} = 6\pi$$

$$\Downarrow$$

$$A + B = 3\pi$$

Solution - 2:

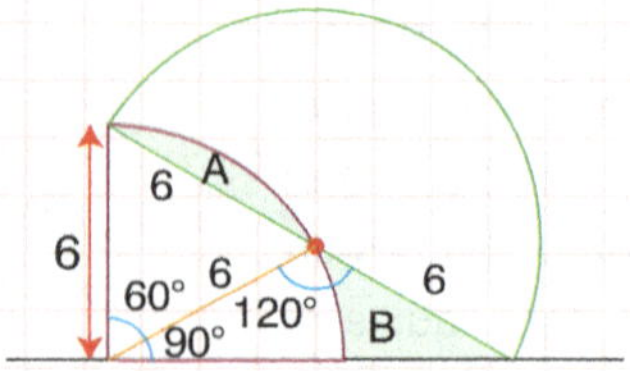

$A = \frac{36\pi}{6} - \frac{36\sqrt{3}}{4}$

$A = 6\pi - 9\sqrt{3}$

$B = \frac{36.\sin 120}{2} - \frac{36\pi}{12}$

$B = 9\sqrt{3} - 3\pi$

$A + B = 3\pi$

Practice 4

Adapted from a geometric puzzle originally shared by Catriona Agg (@Cshearer41) on X

These two circles have the same center. What's the shaded area?

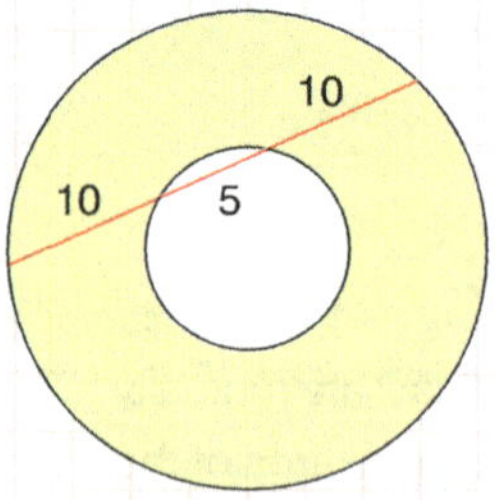

Solution:

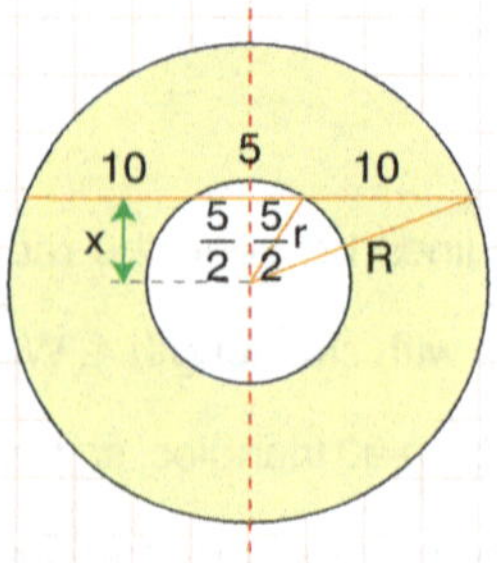

$R^2 = (12 \cdot 5)^2 + x^2$ AND $r^2 = (2 \cdot 5)^2 + x^2$

So: $R^2 - r^2 = \left(\frac{25}{2}\right)^2 + \cancel{x^2} - \left[\left(\frac{5}{2}\right)^2 + \cancel{x^2}\right] = \frac{625}{4} - \frac{25}{4} = \frac{600}{4} = 150$

Annulus area $= \pi R^2 - \pi r^2 = 150\pi$

Practice 5

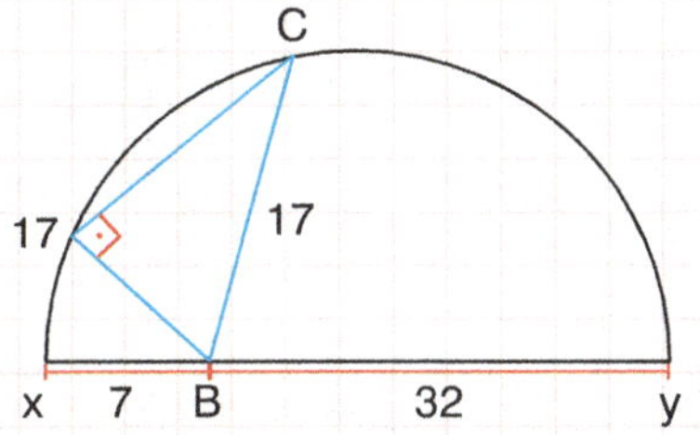

In the figure, XY is the diameter of a semicircle with XY = 39.

Point B lies on $\overline{XY}$ such that XB = 7 and BY = 32.

Points A and C lie on the semicircle (above $\overline{XY}$) so that $\angle BAC = 90°$ and $\overline{BC} = 17$.

Find the area of ΔABC.

Solution:

Let XY = 39 be the diameter, XB = 7, BY = 32, BC = 17.

Let line BA meet the circle again at E.

- Since $\angle BAC = 90°$ and AB and AE are opposite rays, $\angle CAE = 90°$. Thus CE is a diameter, so CE = 39.

Right triangles:

$$AB^2 + AC^2 = 17^2,\ AC^2 + AE^2 = 39^2$$

$$\Rightarrow AE^2 - AB^2 = 39^2 - 17^2 = 1232. \quad (1)$$

Power of point B with chords XY and AE:

$$7 \cdot 32 = BX \cdot BY = BA \cdot BE,$$

and BE = AE − AB, so

$$AB(AE - AB) = 224. \qquad (2)$$

Let = AE − AB. From (2): ABt = 224.

From (1):

$AE^2 - AB^2 = (AE - AB)(AE + AB) = t(t + 2AB) = t^2 + 2ABt = t^2 + 448 = 1232$, so $t^2 = 784$, t = 28, hence AB = 224/28 = 8.

Then

$$AC^2 = 17^2 - 8^2 = 225 \Rightarrow AC = 15.$$

Area:

$$[ABC] = \frac{1}{2} \cdot AB \cdot AC = \frac{1}{2} \cdot 8 \cdot 15 = 60.$$

Practice 6

In the figure shown, are AFC is drawn on isosceles right triangle ABC, with center B. Are AFC is drawn with center D, which is the midpoint of hypotenuse AC. If the length of $\overline{AB}$ is 90 feet, what is the area of the shaded region?

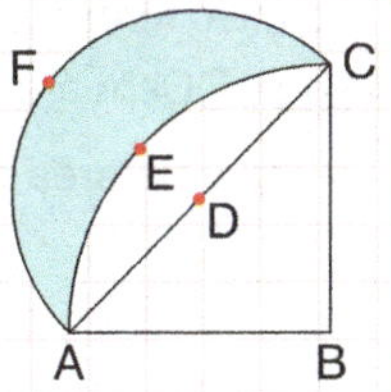

Solution:

Since ABC is an isosceles right triangle with AB = BC = 90,

$$[ABC] = \frac{1}{2} \cdot 90 \cdot 90 = 4050, \quad AC = 90\sqrt{2}.$$

Semicircle AFC: radius $= \frac{AC}{2} = 45\sqrt{2}$,

$$[\text{semicircle AFC}] = \frac{1}{2}\pi(45\sqrt{2})^2 = 2025\pi.$$

Quarter-circle BAEC: radius = 90,

$$[\text{quarter-circle BAEC}] = \frac{1}{4}\pi(90)^2 = 2025\pi.$$

So the shaded area is

$$[ABC] + [AFC] - [BAEC] = 4050 + 2025\pi - 2025\pi = 4050$$

Practice 6

Triangle ABC is inscribed in a circle. The side lengths are

AC = 15, BC = 13, AB = 14.

A diameter through C meets the circle again at E. Find the length of $\overline{CE}$.

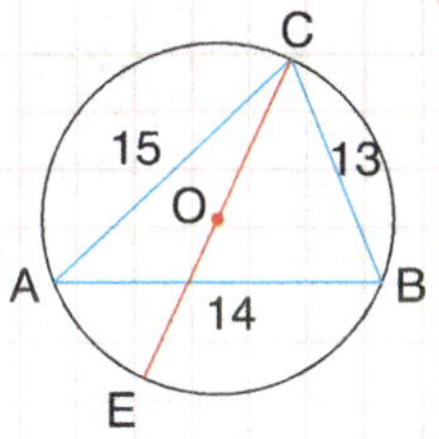

Solution:

Since CE is a diameter, CE = 2R, where R is the circumradius of ΔABC.

First find sin ∠C using the Law of Cosines on side AB:

$$14^2 = 15^2 + 13^2 - 2 \cdot 15 \cdot 13 \cos C$$

$$196 = 225 + 169 - 390 \cos C$$

$$390 \cos C = 398 \Rightarrow \cos C = \frac{33}{65}.$$

Then

$$\sin^2 C = 1 - \cos^2 C = 1 - \left(\frac{33}{65}\right)^2 = \frac{56^2}{65^2} \Rightarrow \text{sic } C = \frac{56}{65}.$$

By the extended Law of Sines,

$$\frac{AB}{\sin C} = 2R \Rightarrow 2R = \frac{14}{56/65} = \frac{14 \cdot 65}{56} = \frac{65}{4},$$

Thus

$$CE = 2R = \frac{65}{4}$$

Practice 8

Adapted from a geometric puzzle originally shared by Catriona Agg (@Cshearer41) on X

What's the area of the large semicircle?

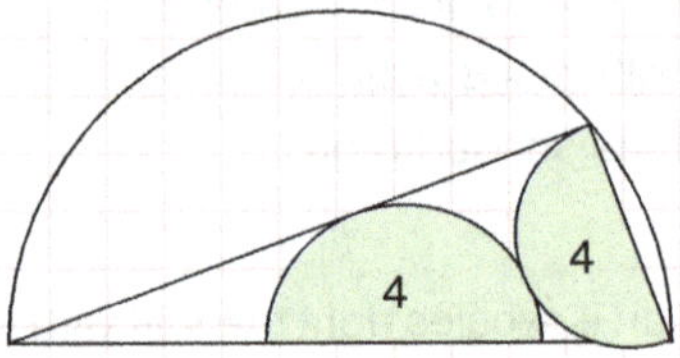

Work Space

Practice 9

Adapted from a geometric puzzle originally shared by Catriona Agg (@Cshearer41) on X

What's the area of the circle?

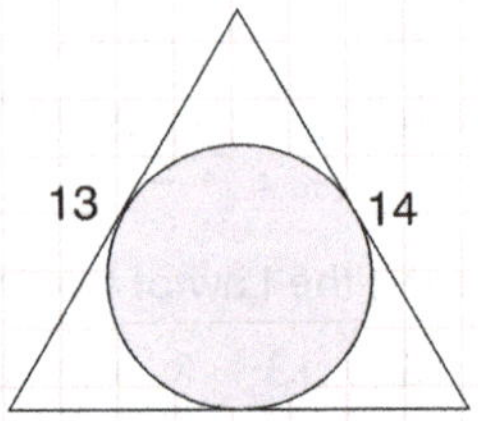

Work Space

Practice 10

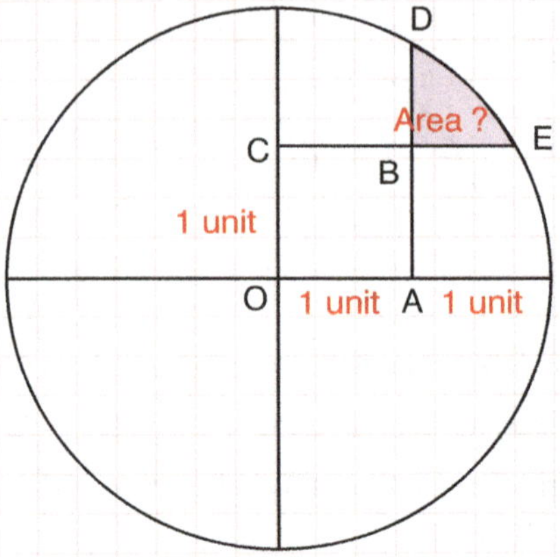

Work Space

MathTopia Press

My Notes:

1.

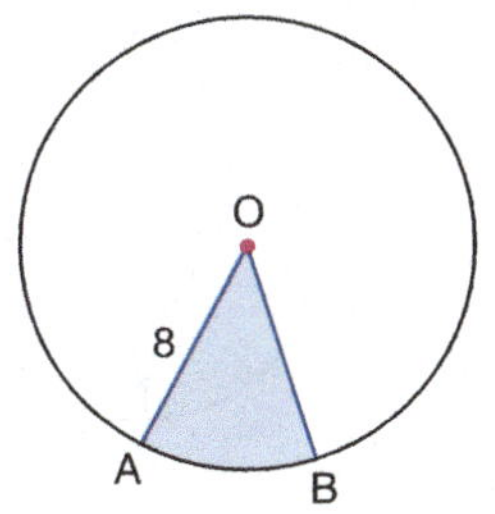

In Circle O, OA = 8 and $m\angle\widehat{AB} = 5\pi$.

What is the area of the shaded section?

2.

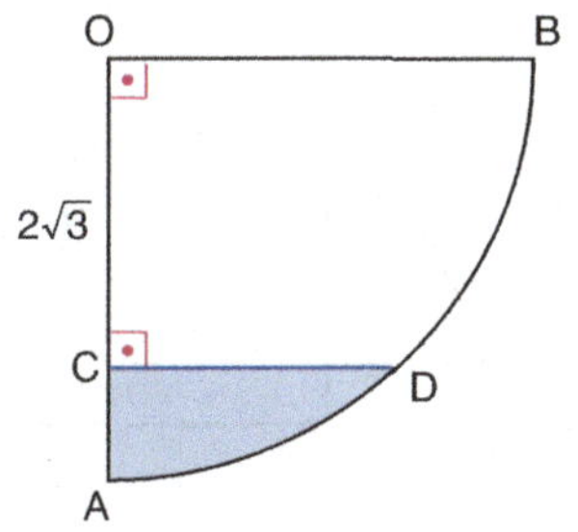

In quarter circle O, CD//OAB, OC = $2\sqrt{3}$, and CD = 2.

What is the area of the shaded region?

3.

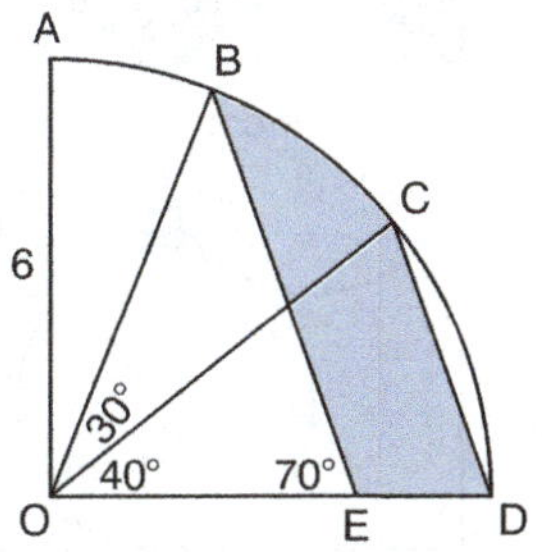

In quarter circle O, $m\angle BOC = 30°$, $m\angle COD = 40°$, and $m\angle OEB = 70°$.

What is the area of the shaded region?

4.

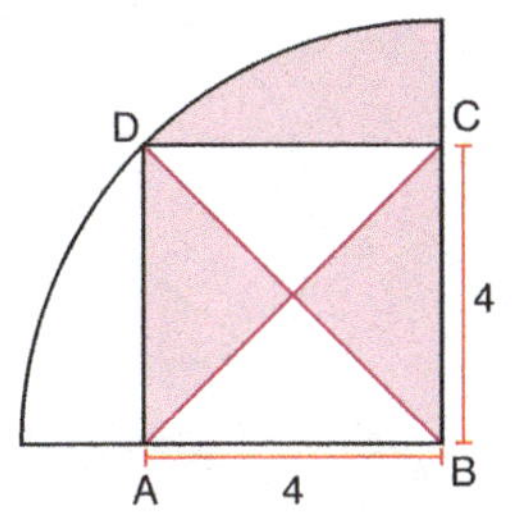

This figure consist of quarter Circle B and Square ABCD.

What is the total area of the shaded regions?

5.

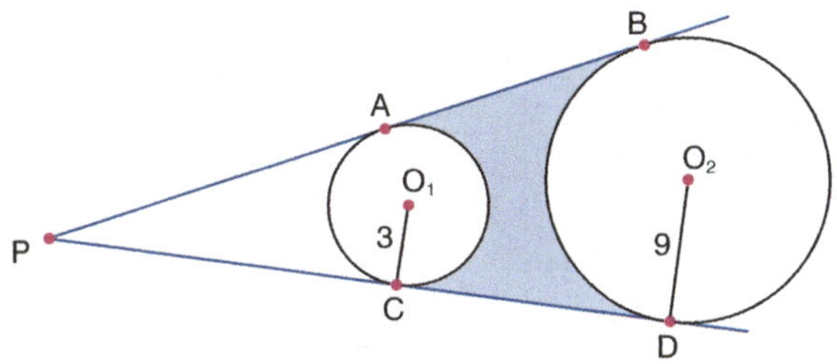

Two circles with radii 9 and 3 are externally tangent.

What is the area of the shaded region?

6.

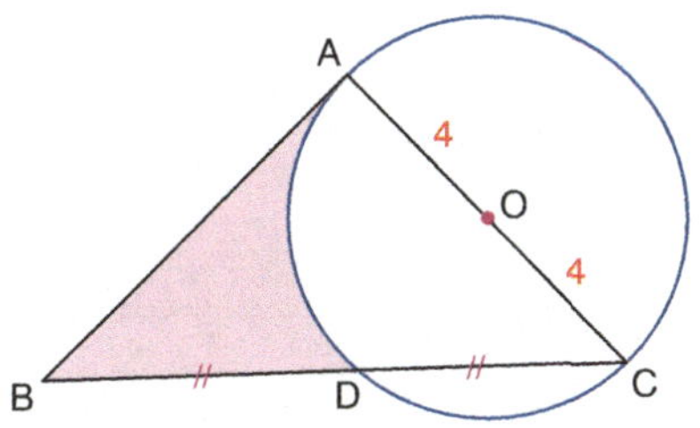

In circle O with diameter of AC = 8, point A is a tangent point and BD = DC.

What is the area of the shaded region?

MathTopia Press

7.

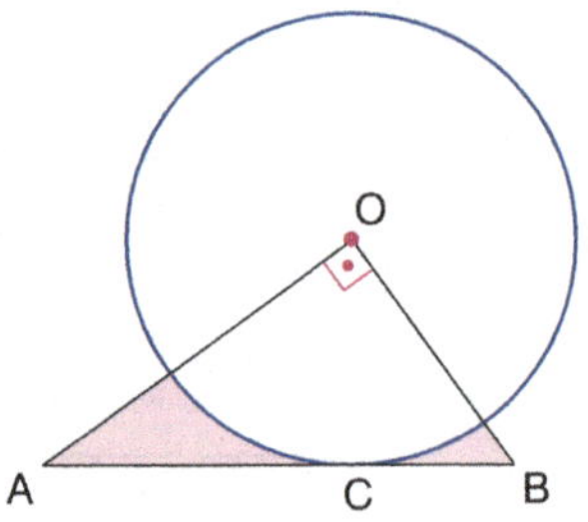

In Circle O, point C is a tangent point and OAB is a right triangle. Given CD = $2\sqrt{3}$ and CB = 2.

What is the area of the shaded region?

8.

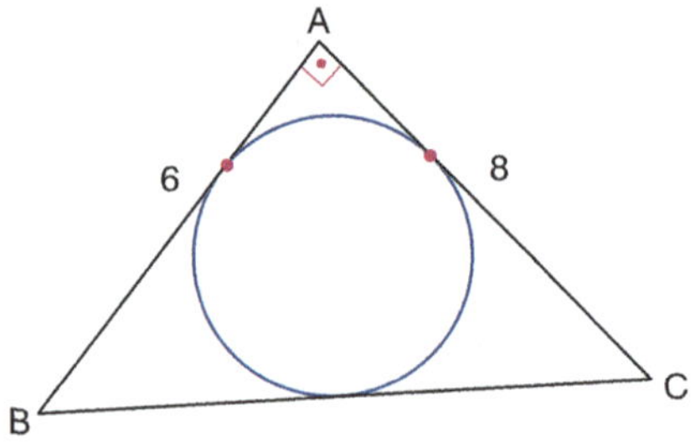

A circle is inscribed in a right triangle. The legs of the triangle are 6 and 8 cm, as shown.

What is the area of the circle?

9. An right isosceles right triangle has one leg of length 8 cm.

What is the area of its inscribed circle?

10.

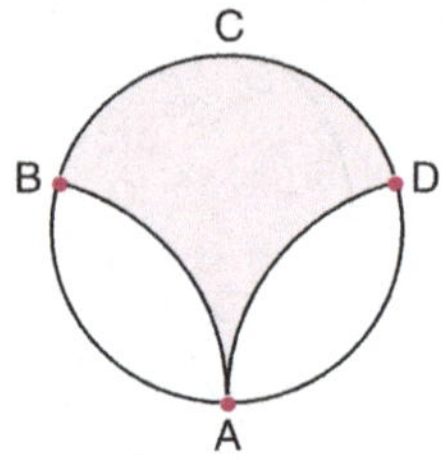

Three circular areas of radius of 8 units bound the region shown. Arcs AB and AD are quarter-circles and arc BCD is a semicircle.

What is the area, in square unit, of the region?

11.

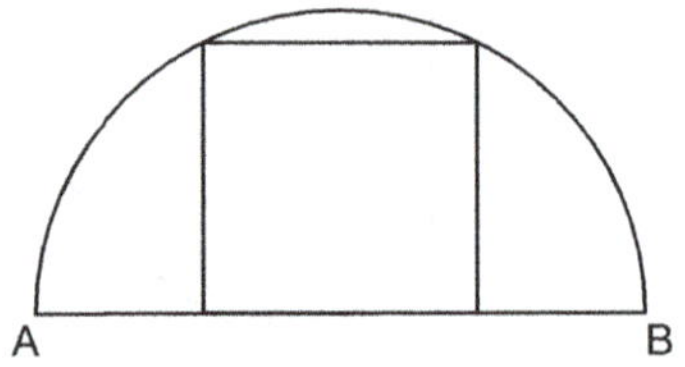

A square of area 40 is inscribed in a semicircle as shown.

What is the area of the semicircle?

12.

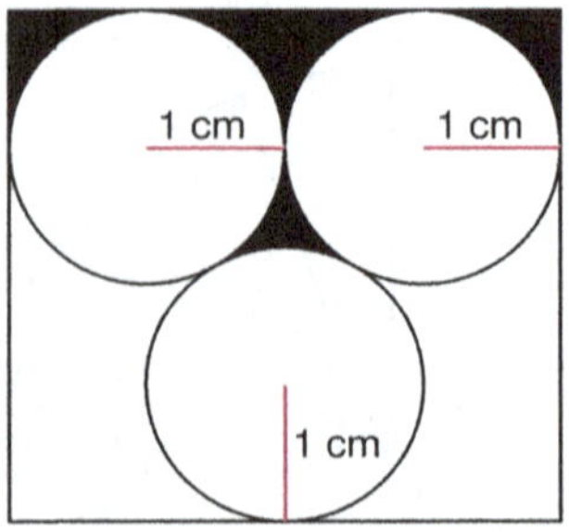

In the figure, three identical circles with radius of 1 replaced in a rectangle as shown.

What is the total area of the shaded parts?

1.

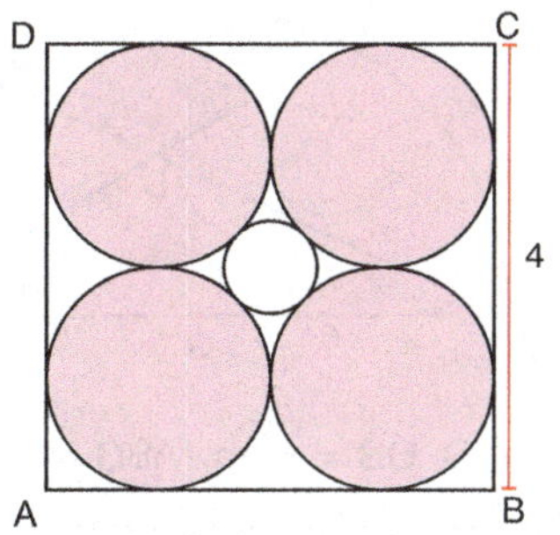

In the figure, ABCD is a square and 4 identical circles touch the small circle as shown.

What is the area of the small circle to the area of the square?

2.

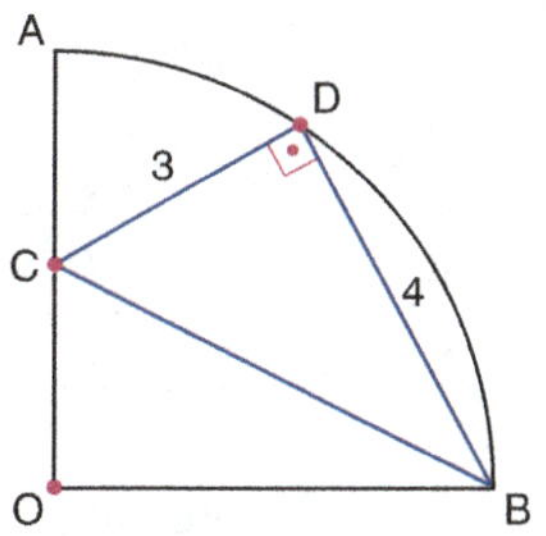

In the figure O is the center of a quarter circle.

$|CD| = 3$, $DB = 4$, and $CD \perp DB$.

Find the area of the quarter circle.

3.

Three circles each have radius 2 cm and pass through centers of the other two circles?

What is the area of the shape that they enclose?

4. Two overlapping circles each have a radius of 4.

If the distance between the centers of two circles is 4, find the area of the overlap region.

5.

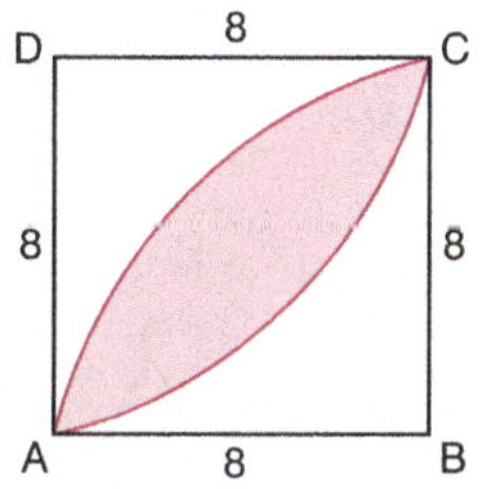

In the figure, ABCD is a square with side lengths 8.

What is the area of the football shape that formed by two quarter circles.?

6.

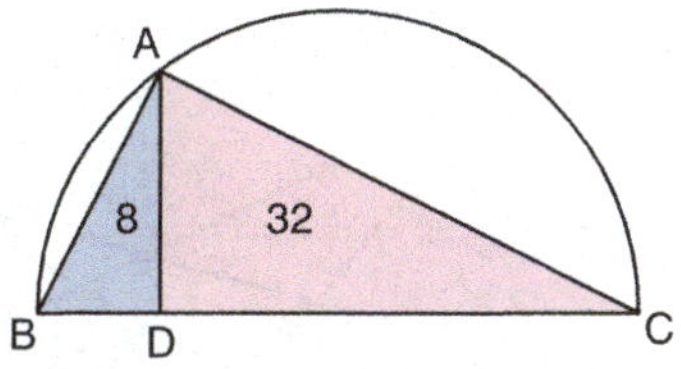

In the figure the area of two right triangles ABD and ADC are 8 and 32, respectively.

What is the area of the semicircle?

7.

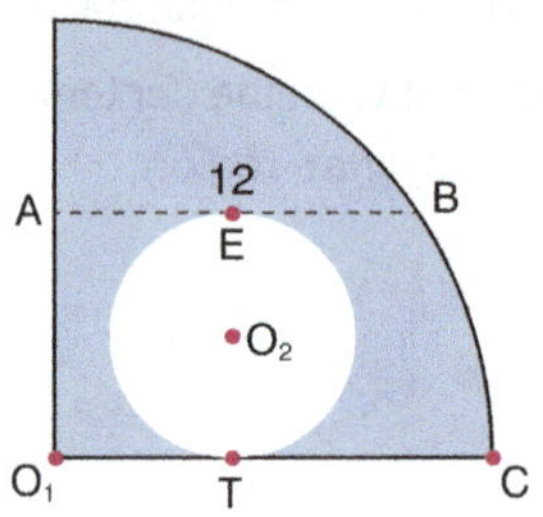

Quarter circle O_1 and circle O_2 are tangent to each other at T. Given that AB = 12 and tangent to circle O_2 at E, find the total shaded area?

8.

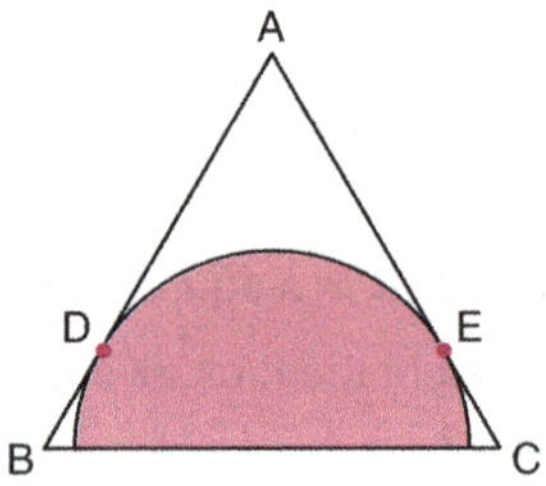

In the figure, a semicircle inscribed in an equilateral triangle and D and E are tangent points. If the side length of the triangle is 2 cm, then find the area of the semi-circle.

9.

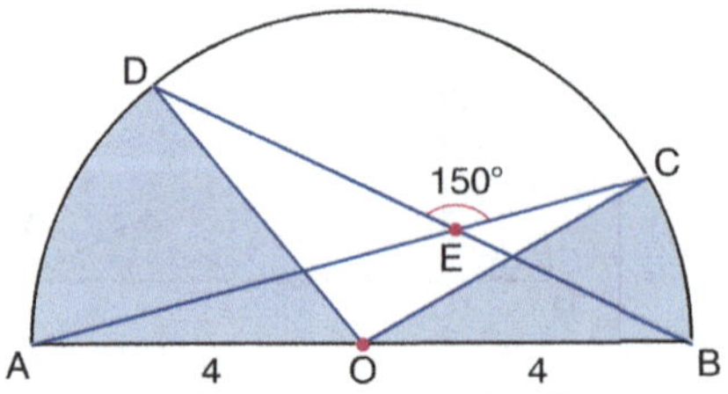

In semicircle O, OB = 4 and m∠DEC 150°.

What is the area of the shaded region?

10.

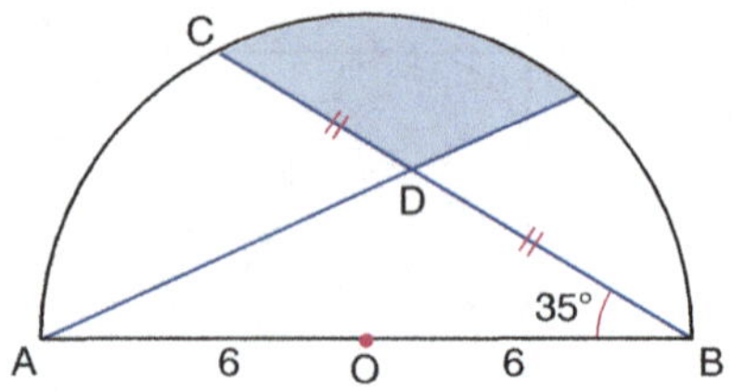

In Semicircle O, OB = 6, m∠ABC = 35°, and CD = DB.

What is the area of the shaded region?

11.

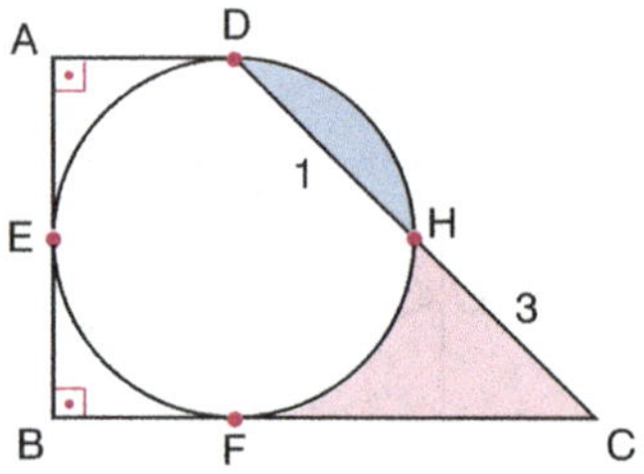

In the figure, D, E, and F are tangent points and ABCD is a trapezoid. If DH = 1 and HE = 3, then find the shaded area?

12.

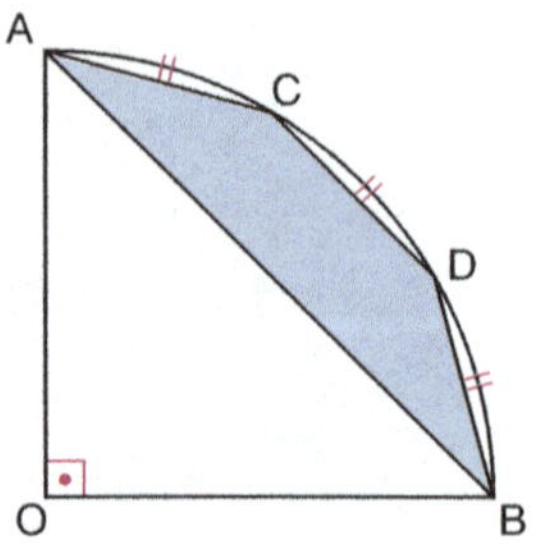

In $\frac{1}{4}$ circle O, AO = 10 and m $\overset{\frown}{AC}$ = m $\overset{\frown}{CD}$ = m $\overset{\frown}{DB}$.

What is the area of the shaded region?

MathTopia Press

1. **(2006 AMC 10B, #8)**

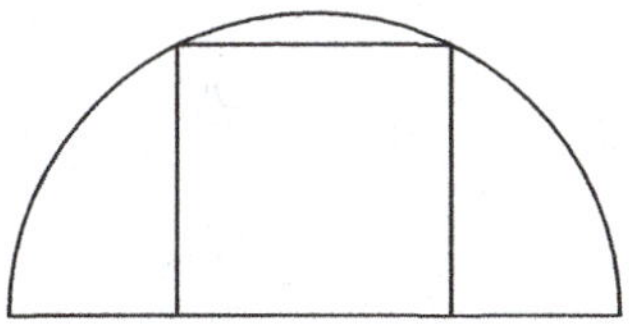

A square of area 40 is inscribed in a semicircle as shown. What is the area of the semicircle?

A) 20π B) 25π C) 30π D) 40π E) 50π

2. **(CWML 14-15 Meet 3, #8)**

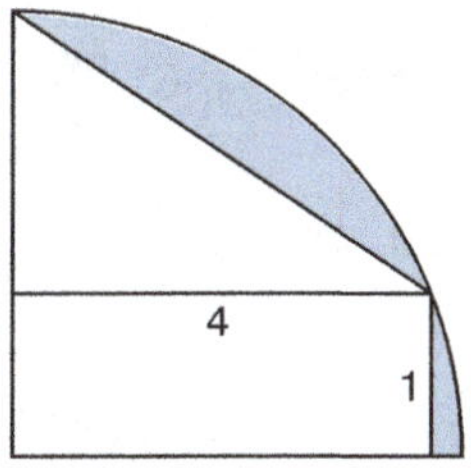

A rectangle with dimensions 4×1 and an adjacent right triangle with base length 4 are enclosed in a quarter circle as shown in the diagram. Find the shaded area.

3. **(CWML 15-16Meet 3, #8)**

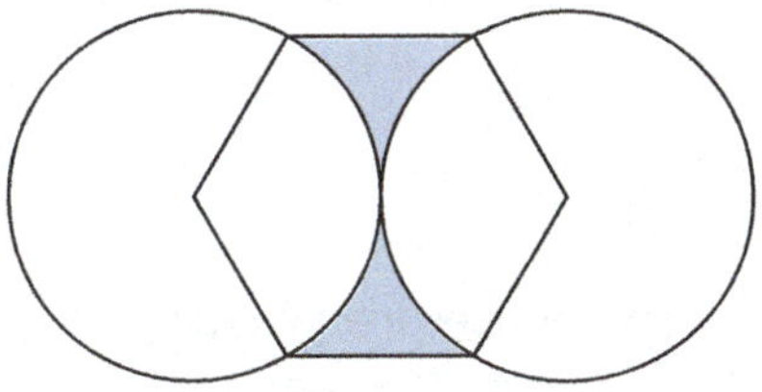

The figure shows a regular hexagon measuring 10 units on each side and two tangent circles each with radius 10 units. Find the area of the shaded region outside the circles and inside the hexagon.

4. **(2002 AMC 12A, #7)**

A $45°$ arc of circle A is equal in length to a $30°$ arc of circle B. What is the ratio of circle A's area and circle B's area?

A) $\frac{4}{9}$ B) $\frac{2}{3}$ C) $\frac{5}{6}$ D) $\frac{3}{2}$ E) $\frac{9}{4}$

5. **(2007 AMC 12A, #10)**

A triangle with side lengths in the ratio 3 : 4 : 5 is inscribed in a circle with radius 3.

What is the area of the triangle?

A) 8.64 B) 12 C) 5π D) 17.28 E) 18

6. **(2002 AMC 10A, #12)**

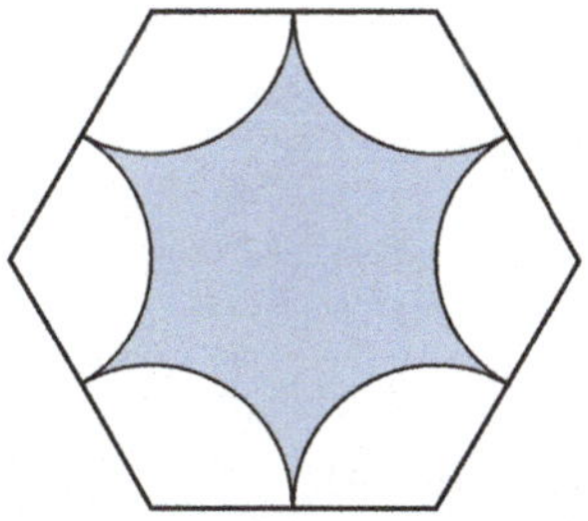

A regular hexagon has side length 6. Congruent arcs with radius 3 are drawn with the center at each of the vertices, creating circular sectors as shown. The region inside the hexagon but outside the sectors is shaded as shown.

What is the area of the shaded region?

A) $23\sqrt{3} - 9\pi$ B) $23\sqrt{3} - 6\pi$ C) $54\sqrt{3} - 18\pi$ D) $54\sqrt{3} - 12\pi$ E) $108\sqrt{3} - 9\pi$

7. (2012 AMC 10B, #16)

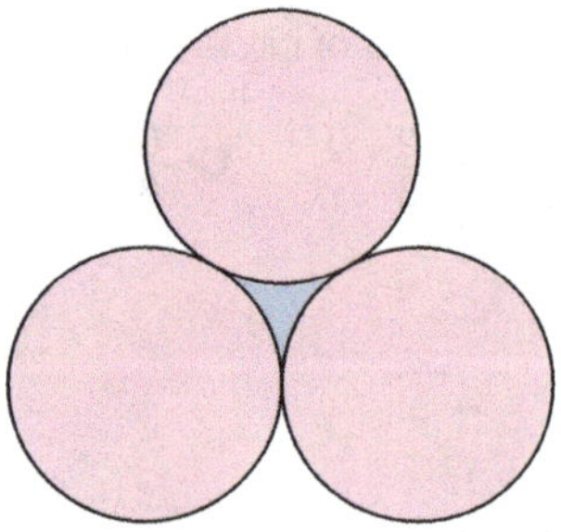

Three circles with radius 2 are mutually tangent. What is the total area of the circles and the region bounded by them, as shown in the figure?

A) $10\pi + 4\sqrt{3}$ B) $13\pi - \sqrt{3}$ C) $12\pi + \sqrt{3}$ D) $10\pi - 9$ E) 13π

8. (2010 AMC 10B, #19)

A circle with center O has area 156π. Triangle ABC is equilateral, $\overline{BC}$ is a chord on the circle OA $= 4\sqrt{3}$, and point O is outside ΔABC.

What is the side length of ΔABC?

A) $2\sqrt{3}$ B) 6 C) $4\sqrt{3}$ D) 12 E) 18

9. (2008 AMC 12A, #13)

Points A and B lie on a circle centered at O, and $\angle AOB = 60°$.

Second circle is internally tangent to the first and tangent to both OA and OB.

What is the ratio of the area of the smaller circle to that of the larger circle?

A) $\frac{1}{16}$ B) $\frac{1}{9}$ C) $\frac{1}{8}$ D) $\frac{1}{6}$ E) $\frac{1}{4}$

10. (2006 AMC 10A, #16)

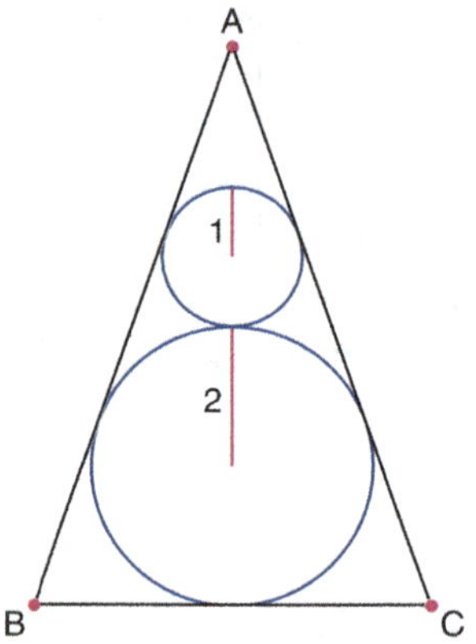

A circle of radius 1 is tangent to a circle of radius 2. The sides of ΔABC are tangent to the circles as shown, and the sides $\overline{AB}$ and $\overline{AC}$ are congruent.

What is the area of ΔABC.

A) 35/2 B) $15\sqrt{2}$ C) $\frac{64}{3}$ D) $16\sqrt{2}$ E) 24

11. (2006 AMC 10B, #24)

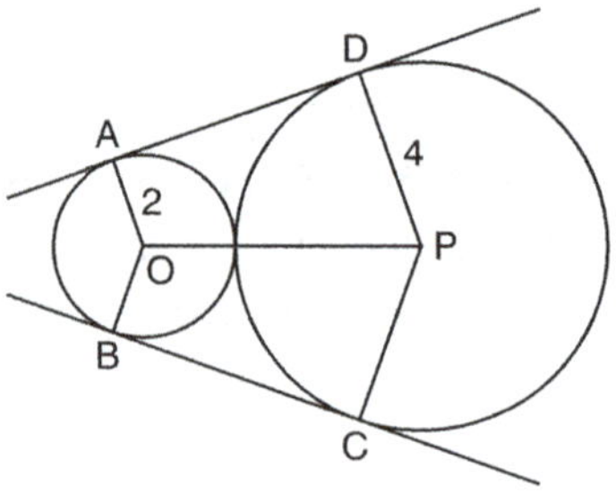

Circles with centers O and P have radii and, respectively, and are externally tangent. Points A and B on the circle with center O and points C and D on the circle with center P are such that AD and BC are common external tangents to the circles.

What is the area of the concavehexagon AOBCPD?

A) $18\sqrt{3}$ B) $24\sqrt{2}$ C) 36 D) $24\sqrt{3}$ E) $32\sqrt{2}$

12. (PiMC 2019 Ind, #25)

ΔABC is an isosceles triangle with AB = BC = 1 and $\angle B = 120°$.

Let O be the circumcenter of ΔABC. Let D be the point on circumcircle of ABC which is diametrically opposite B. Let E be the intersection point of lines AB and CD, and let F be the intersection point of lines AD and BC.

Find the square of the area of ΔOEF.

1.

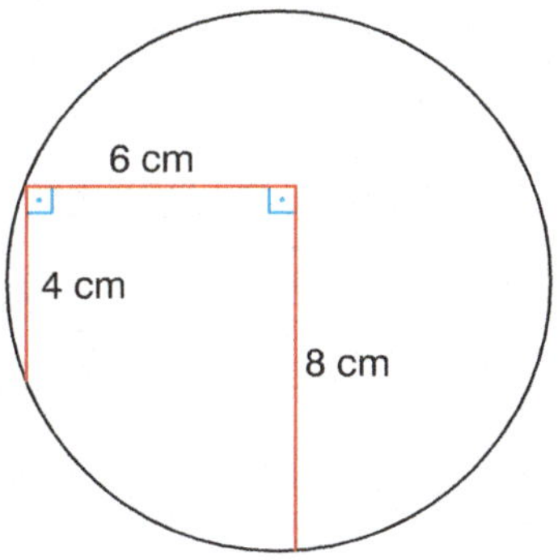

Area of the circle?

2.

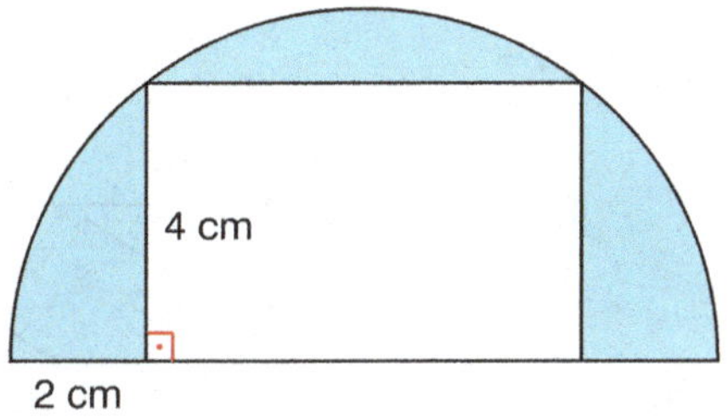

Shaded area?

3.

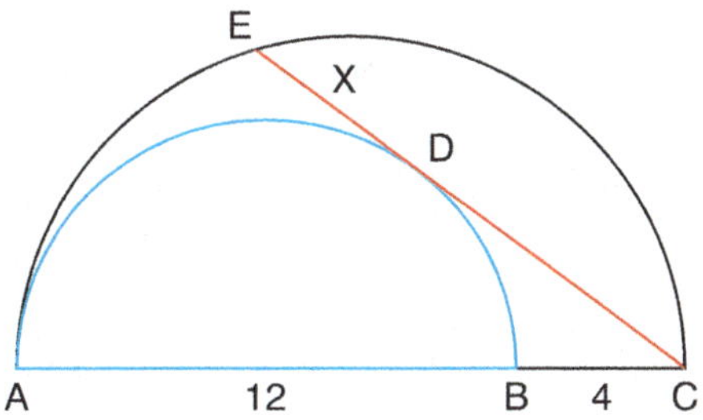

4.

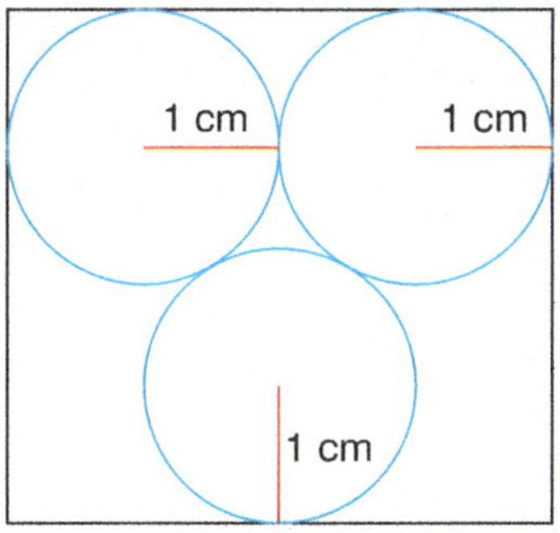

Area of rectangle = ?

5.

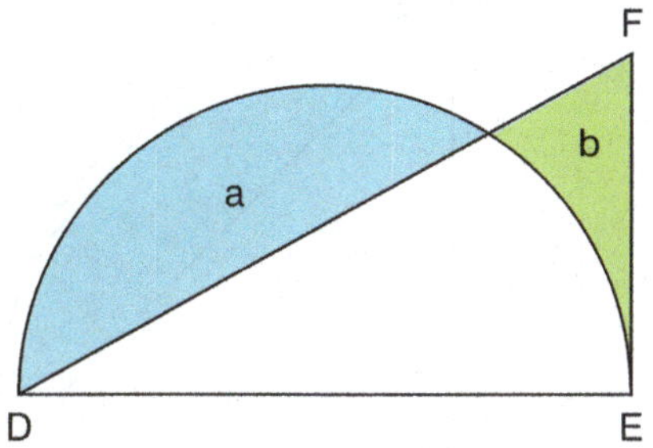

a = b + 47, DE = 20.

EF = ?

6.

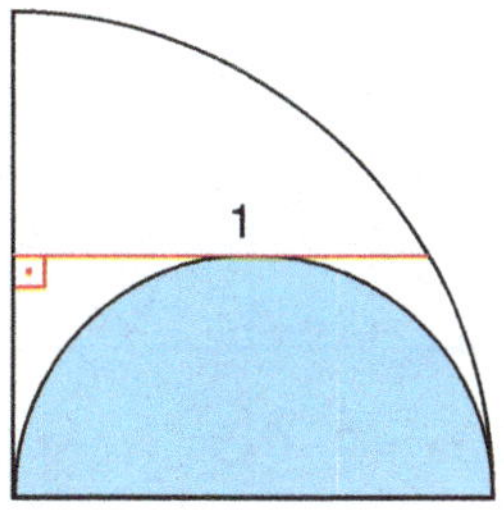

Shaded area = ?

7.

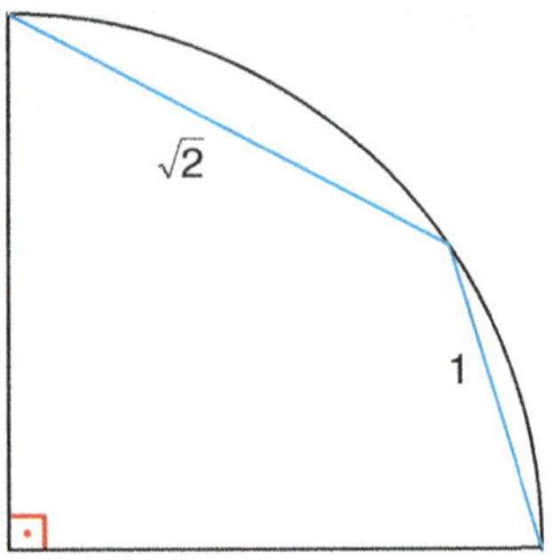

Quarter circle area = ?

8.

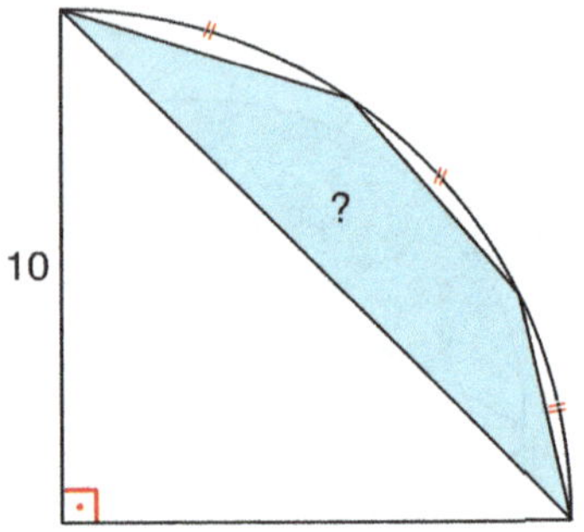

9.

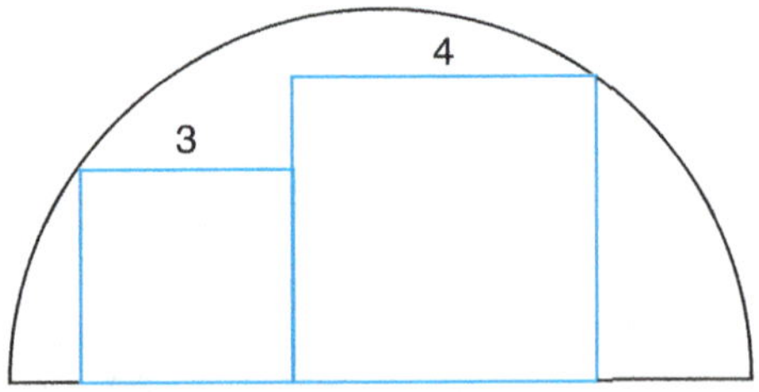

Semicircle area = ?

10.

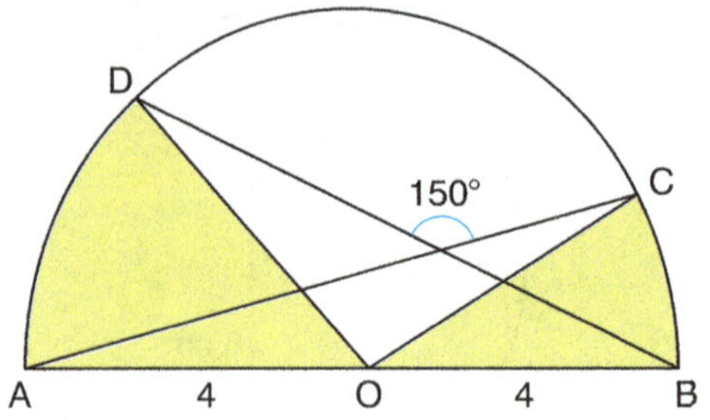

Shaded region = ?

11.

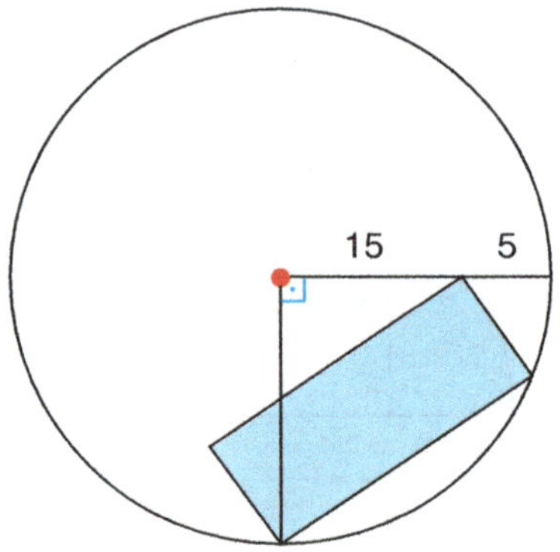

Shaded area = ?

MathTopia Press

12.

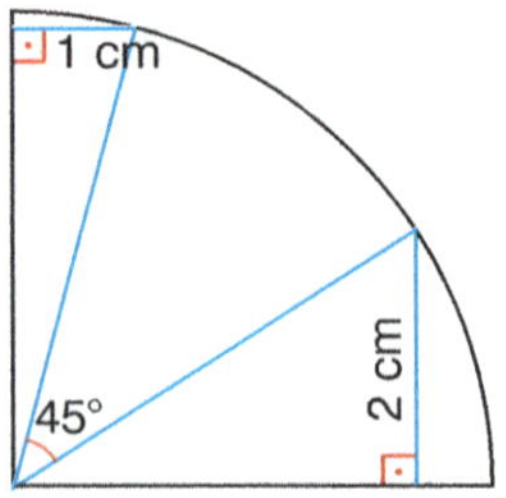

Quarter circle area = ?

CHAPTER 4

CIRCLE - ADVANCED

Theorem 1.1

Ptolemy's Theorem

For a cyclic quadrilateral ABCD (in this order),

$$AC \cdot BD = AB \cdot CD + AD \cdot BC.$$

Proof

Let ABCD be cyclic. Construct a point X on line AD such that $\angle CBX = \angle BAD$. Because equal inscribed angles intercept equal arcs, this construction ensures

$$\angle CBX = \angle CAD \text{ and } BCX = \angle CDA.$$

Thus $\Delta CBX \approx \Delta CAD$ (two angles).

From the similarity: $\dfrac{CB}{CA} = \dfrac{CX}{CD} = \dfrac{BX}{AD}.$ (1)

Cross-multiplying the firs and third ratios of (1) gives

$$CB \cdot AD = BX \cdot CA. \qquad (2)$$

Likewise, comparing the firs and second ratios:

$$CB \cdot CD = CX \cdot CA. \qquad (3)$$

Adding (2) and (3):

$CB \cdot (AD + CD) = (BX + CX) \cdot CA = BA \cdot CA$ since $BX + CX = BC$.

Rearrange:

$$AB \cdot CA = CB \cdot AD + CB \cdot CD = CB(AD + CD).$$

Now perform the same construction but starting from vertex D (or symmetrically exchange roles of diagonals) to get the corresponding relation with BD. Chaining the two symmetric relations (or repeating the argument with roles of A and B interchanged) leads to

$$AC \cdot BD = AB \cdot CD + AD \cdot BC.$$

Theorem 1.2

Power of a Point (tangent from)

Let a circle with center O and radius r be given, and P a point outside the circle. Let PT be a tangent from P to the circle touching at T. Then

$$PT^2 = PO^2 - r^2.$$

Proof

The radius OT is perpendicular to the tangent at T, so ΔPOT is right at T. By the pythagorean theorem,

$$PO^2 = PT^2 + OT^3 = PT^2 + r^2 \Rightarrow PT^2 = PO^2 - r^2.$$

Theorem 1.3

Area of a Triangle via Circumradius

For ΔABC with side lengths a, b, c and circumradius R,

$$K = \frac{abc}{4R}$$

Proof

Use the area formula with an included angle:

$$K = \frac{1}{2}ab \sin C.$$

By the extended law of sines, $a = 2R \sin A$, $b = 2R\sin B$, $c = 2R\sin C$. In particular,

$$\sin C = \frac{c}{2R}.$$

Therefore

$$K = \frac{1}{2}ab \cdot \frac{c}{2R} = \frac{abc}{4R}.$$

Theorem 1.4

Area of a Cyclic Quadrilateral (Brahmagupta)

For a cyclic quadrilateral ABCD with consecutive side lengths $a = AB$, $b = BC$, $c = CD$, $d = DA$ and semiperimeter

$$s = \frac{a+b+c+d}{2},$$

its area is

$$K = \sqrt{(s-a)(s-b)(s-c)(s-d)}$$

Proof

Let the diagonals intersect at $AC = p$.

The area of the quadrilateral equals the sum of the areas of triangles ABC and CDA:

$$K = \frac{1}{2}ab\sin B + \frac{1}{2}cd\sin D.$$

Because the quadrilateral is cyclic, the opposite angles satisfy $B + D = 180°$, so $\sin D = \sin B$. Hence,

$$K = \frac{1}{2}\sin B\,(ab + cd).$$

Now apply the law of cosines to both triangles:

$$\cos B = \frac{a^2+b^2-p^2}{2ab} \text{ and } \cos D = \frac{c^2+d^2-p^2}{2cd}.$$

Since $\cos D = -\cos B$ (because $B + D = 180°$), we can eliminate p from these equations and express $\sin B$ in terms of a, b, c, d. Substituting this value of $\sin B$ into $K = \frac{1}{2}\sin B\,(ab + cd)$ and simplifying yields the desired result:

$$K = \sqrt{(s-a)(s-b)(s-c)(s-d)}..$$

My Notes:

Example 1.1

A cyclic quadrilateral ABCD is inscribed in a circle. The rays AB and DC meet at P outside the circle. Given $PB = 9$, $PC = 6$, $CD = 9$, $AD = 20$, find the area of ABCD.

Solution:

Power of a point at P:

$PB \cdot PA = PC \cdot PD$, $PD = PC + CD = 6 + 9 = 15$

$9 \cdot PA = 6 \cdot 15 \Rightarrow PA = 10 \Rightarrow AB = PA - PB = 1.$

Since ABCD is cyclic, $\Delta PBC \sim \Delta PDA$ and

$$\frac{BC}{AD} = \frac{PB}{PD} = \frac{9}{15} = \frac{3}{5} \Rightarrow BC = \frac{3}{5}\cdot 20 = 12.$$

Now $AB = 1$, $BC = 12$, $CD = 9$, $AD = 20$. Semiperimeter:

$$s = \frac{1+12+9+20}{2} = 21.$$

Brahmagupta:

$$K = \sqrt{(s-a)(s-b)(s-c)(s-d)} = \sqrt{(20)(9)(12)(1)} = \sqrt{2160}$$

$$= 12\sqrt{15}.\ K = 12\sqrt{15}.$$

Theorem 1.5

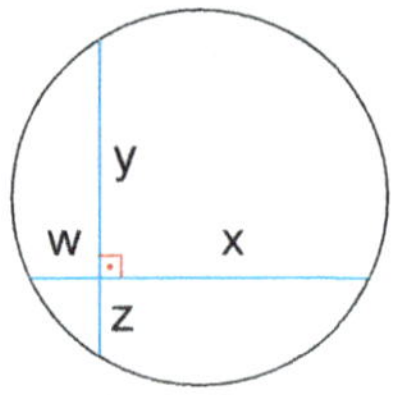

$$4r^2 = w^2 + x^2 + y^2 + z^2$$

Proof

From the geometry of the circle, the radius r forms a right triangle with legs

$$\frac{x-w}{2} \text{ and } \frac{y+z}{2}.$$

By the pythagorean theorem:

$$r^2 = \left(\frac{x-w}{2}\right)^2 + \left(\frac{y+z}{2}\right)^2.$$

Expanding and simplifying:

$$r^2 = \frac{x^2-2xw+w^2+y^2+2yz+z^2}{4} = \frac{w^2+x^2+y^2+z^2}{4}.$$

Multiplying both sides by 4 gives:

$$4r^2 = w^2+x^2+y^2+z^2.$$

Property 1.1

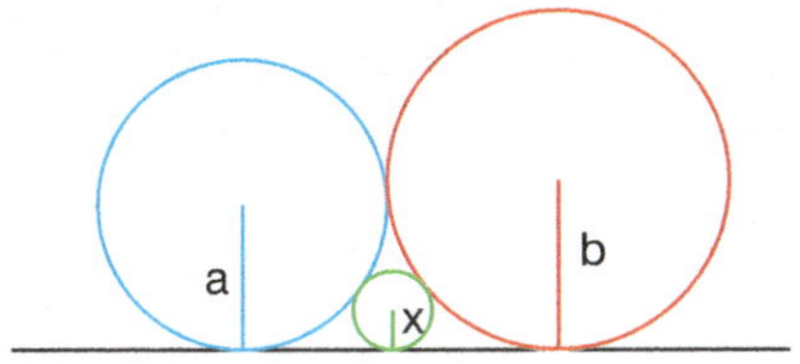

$$\frac{1}{\sqrt{x}} = \frac{1}{\sqrt{a}} + \frac{1}{\sqrt{b}}$$

Property 1.2

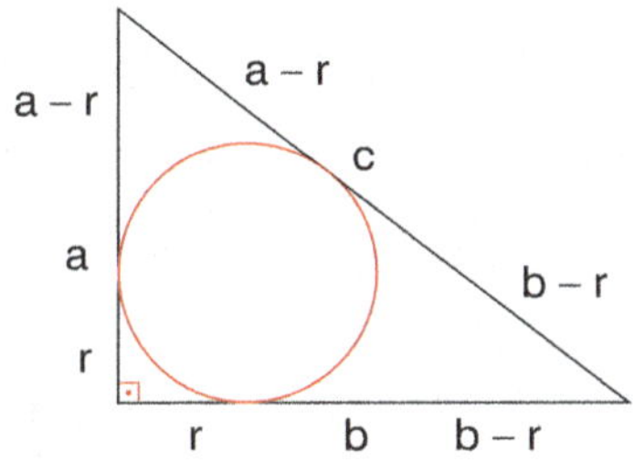

$$A(ABC) = \frac{ab}{2} = \frac{(a+b+c)}{2} r \qquad r = \frac{ab}{a+b+c}$$

$$c = a + b - 2r \qquad r = \frac{a+b-c}{2}$$

Hence, the formula for the inradius in a right triangle not only gives:

$$r = \frac{a+b-c}{2}$$

but also leads directly to the pythagorean theorem:

$$a^2 + b^2 = c^2.$$

Theorem 1.6

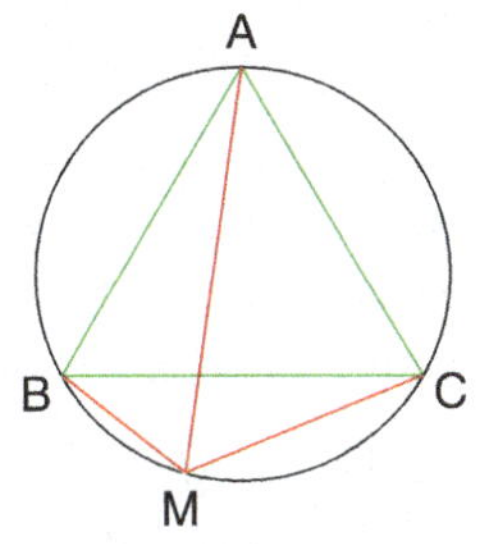

Let ΔABC be an equilateral triangle, and let M be any point on the minor arc BC of its circumcircle.

Then the following relationship holds: $|MA| = |MB| + |MC|$

Proof

Since A, B, C, M lie on the same circle, quadrilateral ABMC is cyclic. We have $\angle BMC = 120°$, $\angle AB = \angle AC = 120°$.

Thus, $\angle BMA = \angle CMA = 60°$

Now, on segment MA, mark a point P such that $|MB| = |MP|$. Because ΔBMP is equilateral, each of its angles is 60°.

In triangle APB and CMB:

$\angle CMB = \angle APM = 120°$, $\angle BCM = \angle BAP$, $|BM| = |BP|$.

Hence, the two triangles are congruent:

$$\Delta CMB \cong \Delta APB.$$

It follows that $|MC| = |PA|$. Finally,

$|MA| = |MP| + |PA| = |MB| + |MC|$.

$$|MA| = |MB| + |MC|.$$

Theorem 1.7

In equilateral ΔABC, a line through A meets BC at D and the circumcircle again at P.

Prove $\frac{1}{PD} = \frac{1}{PB} + \frac{1}{PC}$.

Proof

1. **Equal arc**

   Since AB = AC, arc AB and AC are equal, so

   $$\angle BPA = \angle CPA.$$

   Thus, AP bisects $\angle BPC$.

2. **Angle bisector in** ΔPBC

   $$\frac{BD}{DC} = \frac{PB}{PC}. \qquad (1)$$

3. **Power of a point** D

   $$DA \cdot DP = DB \cdot DC. \qquad (2)$$

4. **Ptolemy for cyclic** ABPC

   $$AB \cdot PC + AC \cdot PB = AP \cdot BC \Rightarrow AP = PB + PC \qquad (3)$$

5. **Combine results**

   From (1) – (3), we find $DP = \frac{PB \cdot PC}{PB + PC}$.

   Therefore, $\frac{1}{PD} = \frac{1}{PB} + \frac{1}{PC}$.

Example 1.2

A, B and C are points on a circle.

- BC bisects angle ABQ.
- PBQ is a tangent to the circle.

Angle CBQ = x

Prove that AC = BC

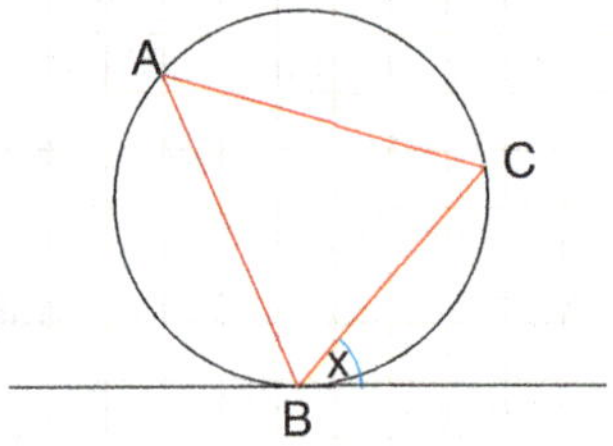

Solution:

- By the tangent-chord theorem, the angle between the tangent BQ and chord BC equals the angle in the opposite arc subtended by BC: $\angle CBQ = CAB$.

  (Given $\angle CBQ = x$, this gives $\angle CAB = x$.)
- Since BC bisects $\angle ABQ$, we have $\angle ABC = \angle CBQ = x$.
- Hence $\angle ABC = \angle CAB$, so triangle ABC is isosceles with AC = BC.

Example 1.3

In triangle ABC, the bisector of angle A intersects side BC at D and the circumcircle again at E.

To Prove:

a. $EB^2 = ED \cdot EA$

b. $AD \cdot AE = b \cdot c$

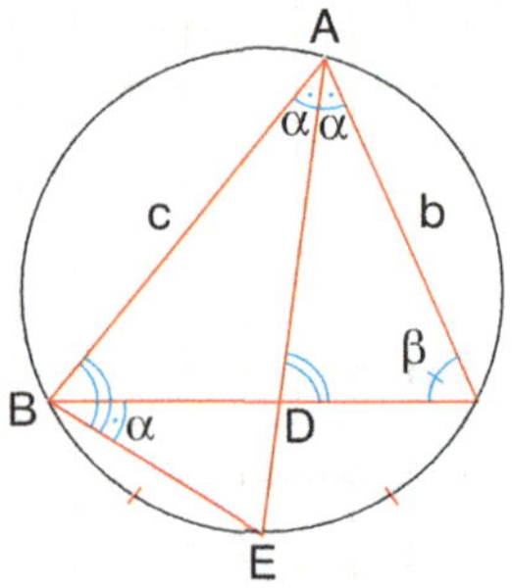

Solution:

a. The bisector of angle A, that is AD, meets the circumcircle again at E, which lies on the arc BEC. Consider triangles EBD and EAB:

- $\angle E$ is common.
- $\angle EBD = \angle EAB$ (they subtend the same arc).
- $\angle EBA = \angle EDB$.

Therefore, by AAA similarity, $\Delta EBD \sim \Delta EAB$.

Hence, $\frac{EB}{EA} = \frac{ED}{EB} \Rightarrow EB^2 = ED \cdot EA$.

b. Similarly, triangles ABE and ADC are similar (AAA), so

$\frac{AB}{AD} = \frac{AE}{AC}$.

Cross-multiplying gives $AD \cdot AE = AB \cdot AC = b \cdot c$.

Thus, $AD \cdot AE = b \cdot c$.

Practice 1

Cyclic quadrilateral ABCD has lengths BC = CD = 3 and DA = 5 with $\angle CDA = 120°$. What is the length of the shorter diagonal of ABCD?

A) $\frac{31}{7}$ B) $\frac{33}{7}$ C) 5 D) $\frac{39}{7}$ E) $\frac{41}{7}$

Solution:

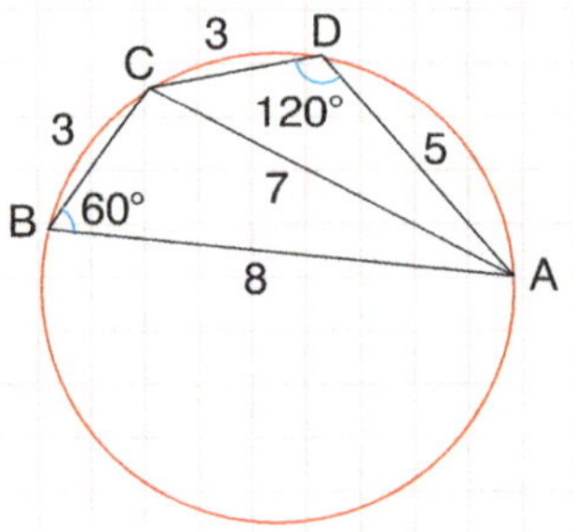

~diagram by erics118

First, $\angle CBA = 60°$ by properties of cyclic quadrilaterals.

Let AC = u. Apply the law of cosines on ΔACD:

$u^2 + 3^2 + 5 - 2(3)(5)\cos 120°$ $u = 7$

Let AB = u. Apply the law of cosines on ΔABC:

$7^2 = 3^2 + u^2 - 2(3)(u)\cos 60°$ $u = \frac{3 \pm 13}{2}$ $u = 8$

By ptolemy's theorem,

$AB \cdot CD + AD \cdot BC = AC \cdot BD$ $8 \cdot 3 + 5 \cdot 3 = 7BD$

$$BD = \frac{39}{7}$$

Practice 2

In a semicircle with diameter $\overline{AB}$ of length 24, points C and D lie on the semicircle with B at the right end of the diameter. The chord $\overline{BC}$ has length 18. If $\angle ABD = 2\angle ABC$, then what is BD?

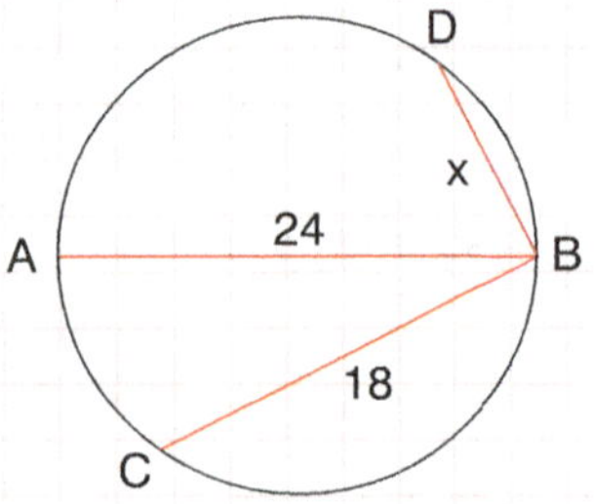

Solution:

Since AB is a diameter, $\angle ACB = 90°$ and

$AC = \sqrt{AB^2 - BC^2} = \sqrt{24^2 - 18^2} = 6\sqrt{7}$.

Let O be the center. The condition $\angle ABD = 2\angle ABC$ means the central angle $\angle BOD$ equals $180° - 4\angle BOC/2$. Dropping perpendiculars from O to chords BD and AC gives two right triangles whose similarity yields the chord relation

$$\frac{BB}{AB} = 1 - 2\left(\frac{AC}{AB}\right)^2.$$

Hence

$$BD = 24\left(1 - 2 \cdot \frac{(6\sqrt{7})^2}{24^2}\right) = 24\left(1 - \frac{14}{8}\right) = 24 \cdot \frac{1}{8} = 3.$$

Practice 3

Adapted from a geometric puzzle originally shared by Catriona Agg (@Cshearer41) on X

What's the area of the rectangle?

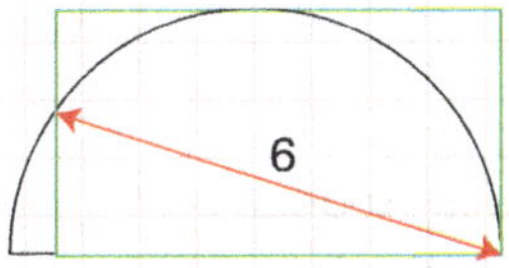

Solution:

- In right Δ formed by radius to P, the horizontal leg is a and the vertical leg is b. Hence $a^2 + b^2 = r^2(1)$.
- The segment PB from P to the right-bottom corner B is given as 6. The horizontal and vertical legs for PD are a + r and b, so $(a + r)^2 + b^2 = 6^2(2)$.

Subtract (1) from (2):

$(a + r)^2 + b^2 - (a^2 + b^2) = 36 - r^2 \Rightarrow 2ar + r^2 = 18.$

Thus $r(a + r) = 18$.

Practice 4

a. If BP = 5 and PC = 20

find AD.

b. If $m\overset{\frown}{BP} : m\overset{\frown}{PC} = 1 : 3$

find the radius of the circle

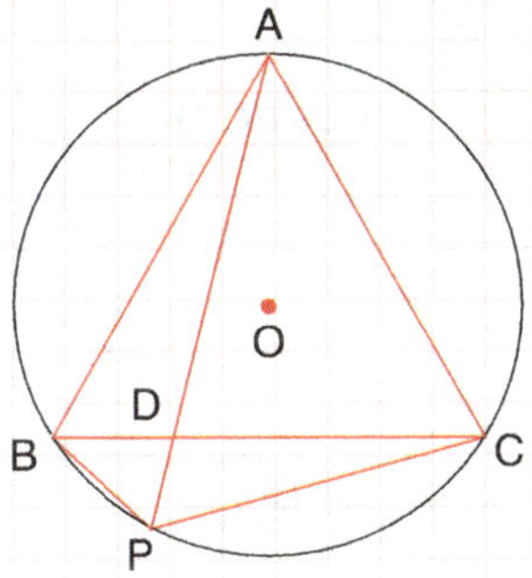

Solution:

a. PB = 5, PC = 20

From vyclic ABPC with AB = AC (equilateral), Ptolemy gives AP = PB + PC = 25.

From the proven relation $\frac{1}{PD} = \frac{1}{PB} + \frac{1}{PC}$

$PD = \frac{PB \cdot PC}{PB + PC} = \frac{5 \cdot 20}{25} = 4.$

Since A – D – P are collinear, AP = AD + DP. Hence

AD = AP – PD = 25 – 4 = 21.

a. $R = 10\sqrt{2}$

Reason: $\overset{\frown}{BP} : \overset{\frown}{PC} = 1 : 3 \Rightarrow \overset{\frown}{PC} = 90°.$

Chord PC subtends 90°, so in ΔPOC (with OP = OC = R) we get $PC = R\sqrt{2}$.

Given PC = 20, hence $R = 20/\sqrt{2} = 10\sqrt{2}$

Practice 5

Triangle ABC has side lengths AB = 11, BC = 24, and CA = 20. The bisector of ∠BAC intersects $\overline{BC}$ in point D, and intersects the circumcircle of ΔABC in point E ≠ A. The circumcircle of ΔBED intersects the line AB in points F ≠ B. What is CF?

A) 28 B) $20\sqrt{2}$ C) 30 D) 32 E) $20\sqrt{3}$

Solution:

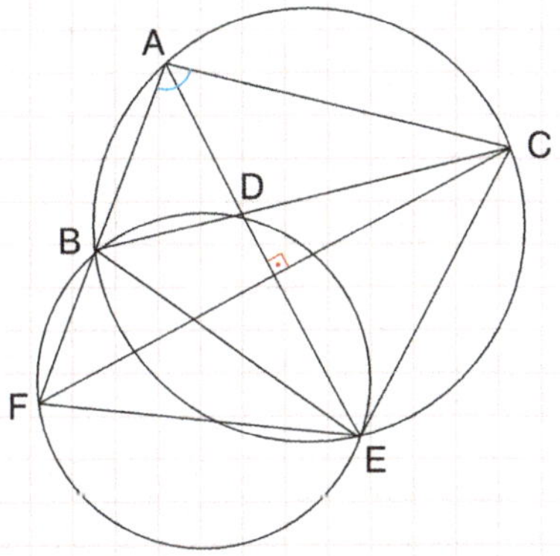

By the law of cosine cos

$$A = \frac{AC^2 + AB^2 - BC^2}{2 \cdot AC \cdot AB} = \frac{20^2 + 11^2 - 24^2}{2 \cdot 20 \cdot 11} = -\frac{1}{8}$$

As ABEC is a cyclic quadrilateral, ∠CEA = ∠CBA. As BDEF is a cyclic quadrilateral, ∠CBA = ∠FEA.

- ∠CEA = ∠FEA and ∠CAE = ∠FAE
- ΔAFE ≅ ΔACE by ASA

Hence, AF = AC = 20

By the law of cosine

$$CF = \sqrt{20^2 + 20^2 - 2 \cdot 20 \cdot 20\left(-\frac{1}{8}\right)} = \sqrt{900} = C30$$

Note that F is C's reflection over line AE, quadrilateral ACEF is a kite symmetrical by line AE, AE ⊥ CF.

Practice 6

In the semicircle shown, AB is the diameter with AB = 12 cm. Points C and D are on the semicircle such that BC = CD = 3 cm. What is 2 · AD, in cm?

Work Space

Practice 7

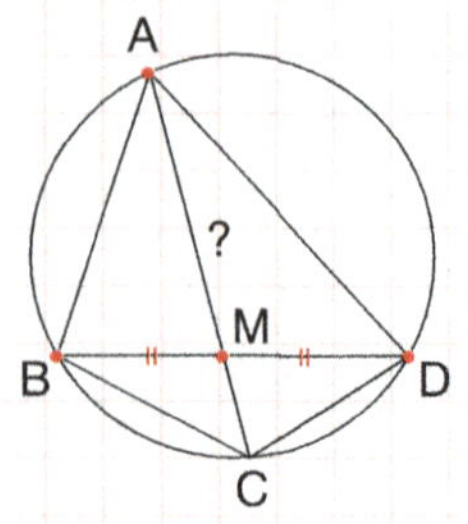

AB – CB = 1

AD + CD = 7

BM = DM

AC = ?

Work Space

Practice 8

Adapted from a geometric puzzle originally shared by Catriona Agg (@Cshearer41) on X

One third of this regular hexagon is shaded. What's the area of the semicircle?

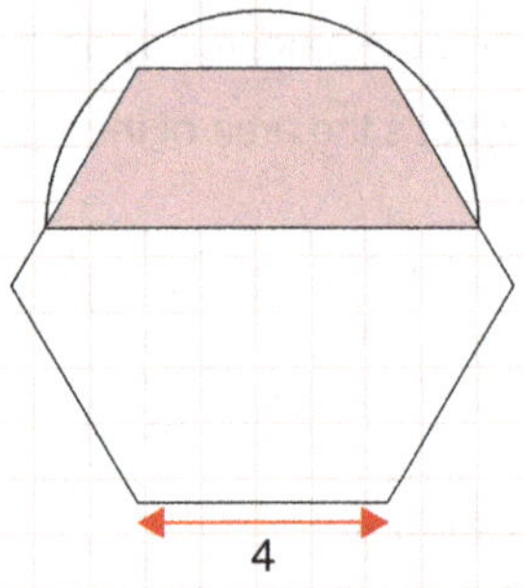

Work Space

Practice 9

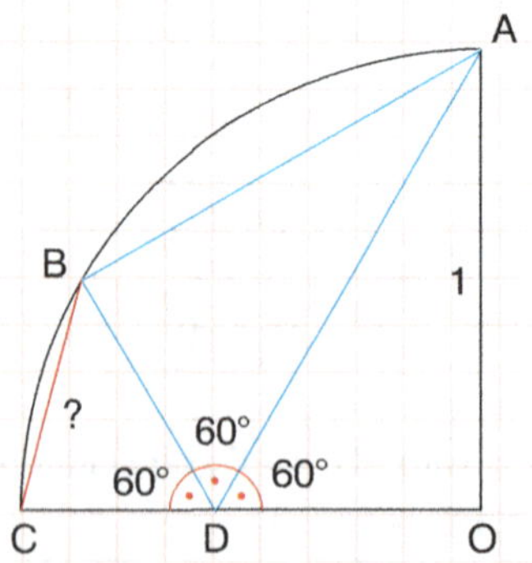

BC = ?

Work Space

MathTopia Press

1. A right triangle with legs of length 3 and 4 inches is given. Find the radius of the inscribed circle.

2. Circle O has diameter AB. Extend AB through B to point P, and draw its tangent to O with intersection point D. Given that PB = 6 and PD = 8, find the radius of circle O.

3. Inside a circle, chords AB and CD intersect at P in the right angles. Given that AP = 8 and BP = 14, and CD = 20, find the radius of the circle.

4. Inside a circle, chords AB and CD intersect at G. Point M is on the segment of GB and the midpoint of AB and point N is on the segment of GD and the midpoint of CD. Given that CG = 8, GN = 5, and GM = 9, find AG.

5. A right triangle has an incircle with a radius of 4 and perimeter of 32, Find the area that is outside of the circle and but inside the triangle.

6.

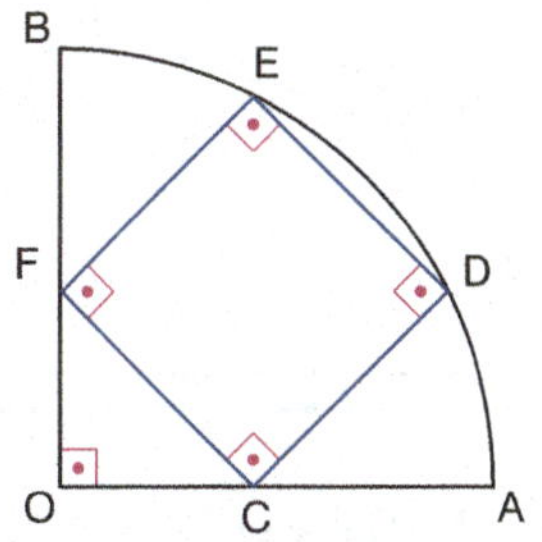

O is a quarter circle and CDEF is a square. If AO = 5 units, then find the perimeter of the square.

7.

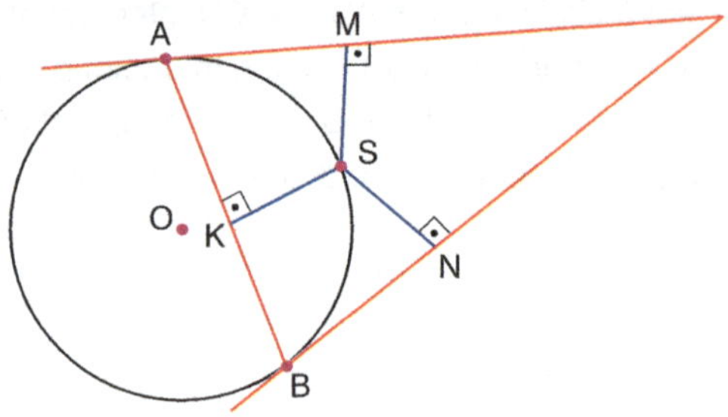

In cirle O, points A and B are tangent to the circle and MS⊥PA, SN⊥BP, and SK⊥AB.

Given that SM = 4 and SN = 9, then find KS.

8.

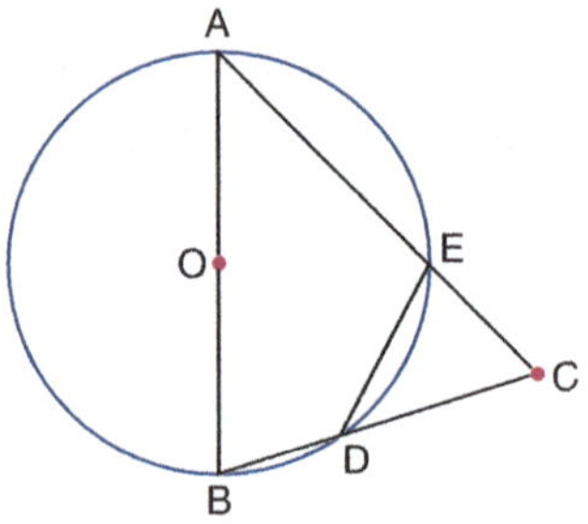

In circle O, AB is a diameter of the circle.

AB = AC and BC = 10.

What is DE?

9.

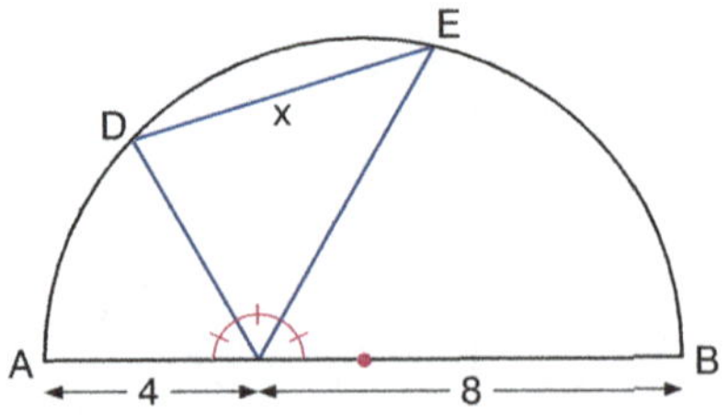

A semicircle with diameter 12. Angles ACD, ECE, and BCE are equal each other.

Given that AC = 4 and CB = 8, find DE.

10.

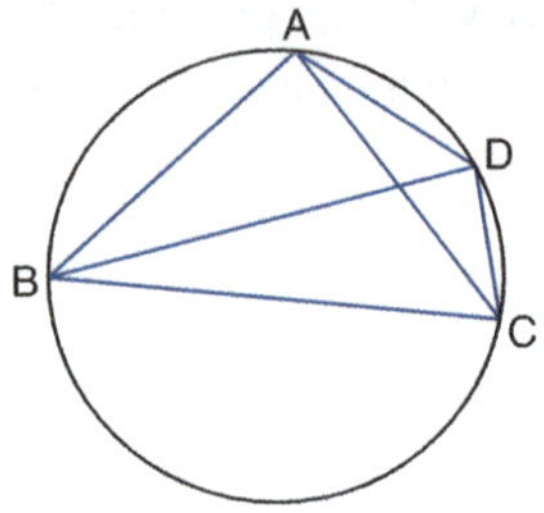

In the figure, ABC is an equilateral triangle.

Given that AD = 5 and DC = 2, find BD.

11.

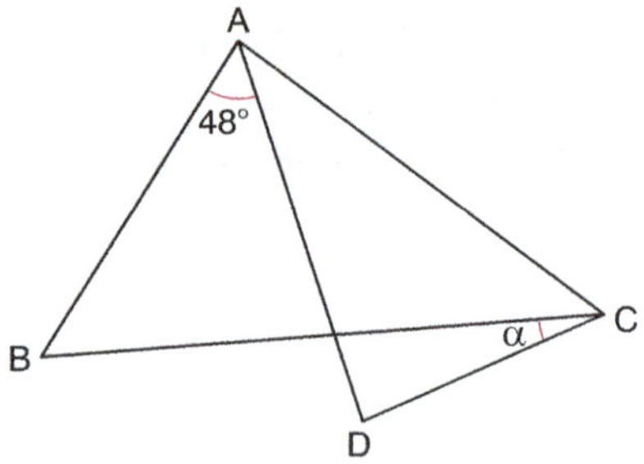

In the figure, AB = AD = AC.

If m∠BAD = 48° then find m∠BCD = α.

12. Two circles are internally tangent (one is within the other). A line passed through points B, C, and the center of the smaller circle and crosses the bigger one at points A and D.

Find the ratio of radii of circles if AB : BC : CD = 2 : 4 : 3

1.

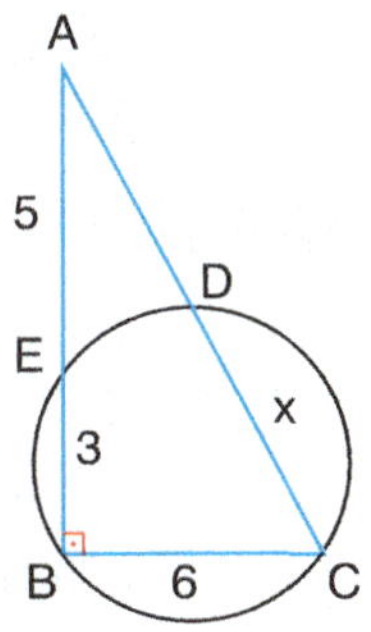

Calculate the length x

2.

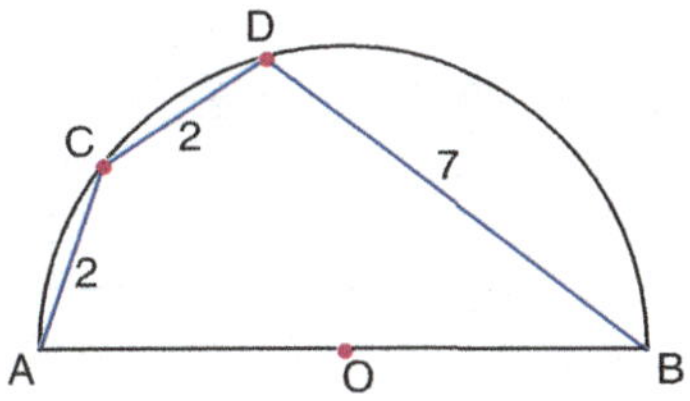

In semicircle O, AC = CD = 2 and BD = 7.

What is the radius of the semicircle?

3.

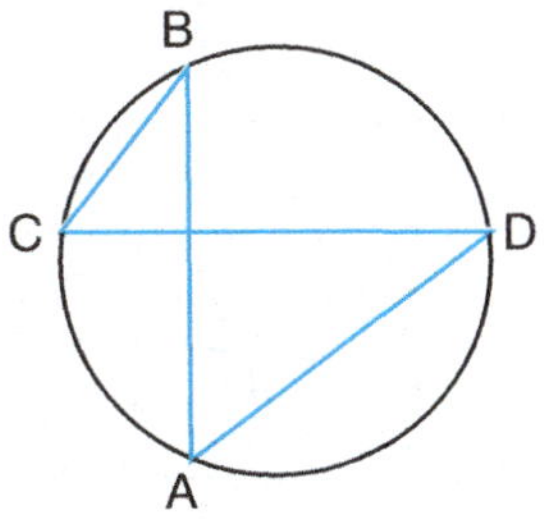

In the figure, AB⊥CD. Given that AD = 40 and CB = 30, find the radius of the circle.

4.

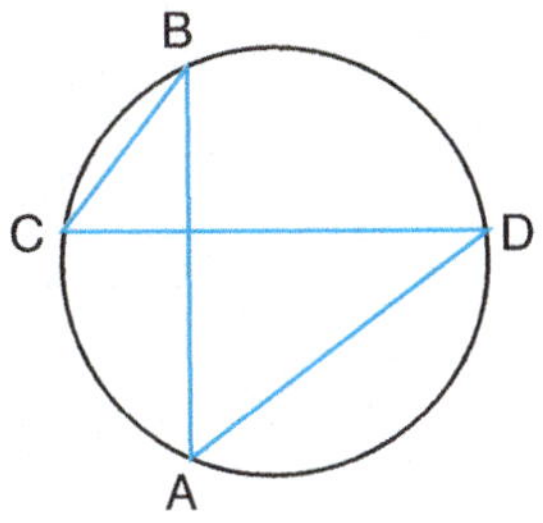

In circle O, m∠ACD = m∠BCE=30.

Given that AC = 12 and CB = 6, find CD – CE.

5.

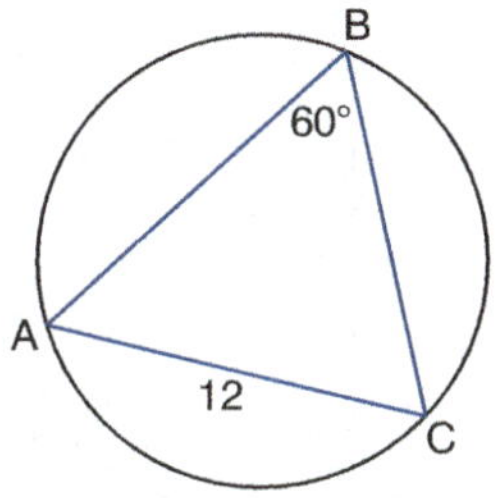

In the figure, A, B, and C are on the circle.

Given that ∠BAC = 60 and AC = 12, find the area of the circle.

6.

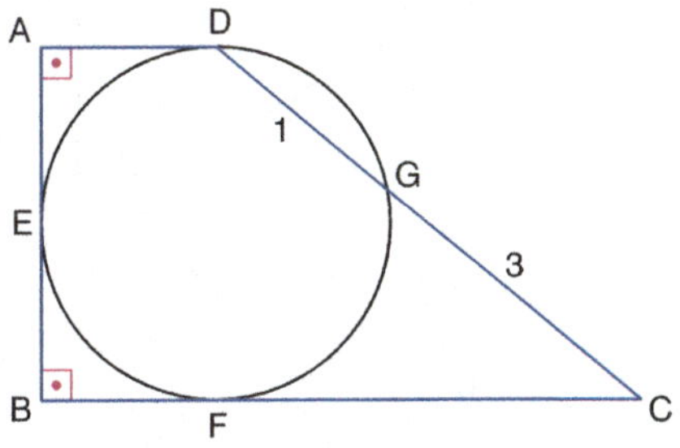

In the figure, points D, E, and F are tangent to the circle.

Given that DG = 1 and GC = 3, find the area of ABCD.

7.

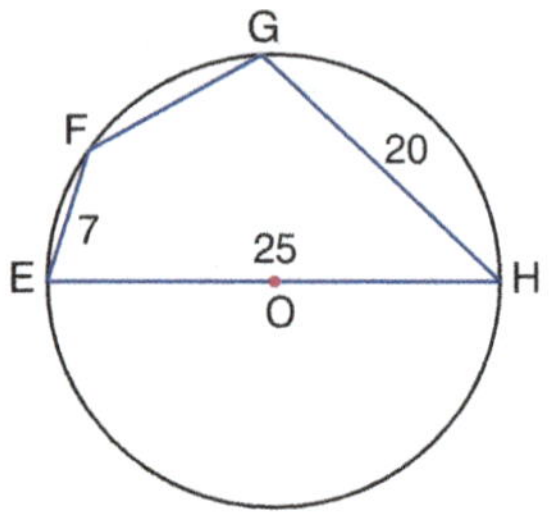

In the figure, EH is a diameter with the length of 25.

Given that EF = 7 and GH = 20, find the perimeter of the inscribed quadrilateral.

8.

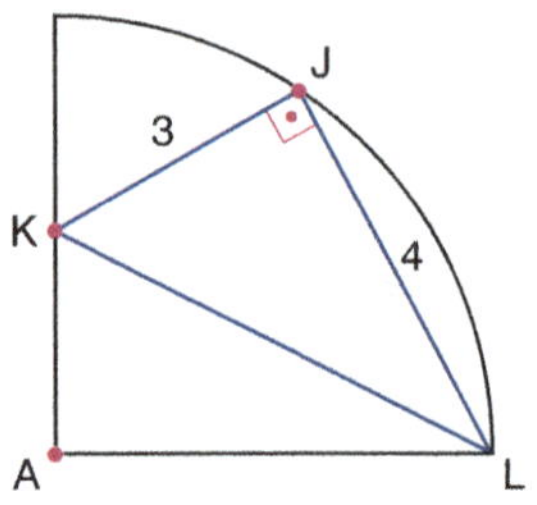

In quarter circle A, KJ⊥JK.

Given that KJ = 3 and JK = 4, find the area of the quarter circle.

9.

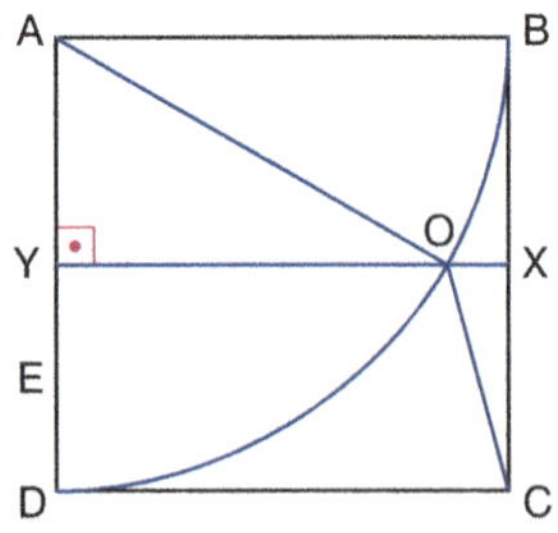

ABCD is a square and quarter circle with radius AD meets OY perpendicular.

Given that AY = YD, then find m∠AOC.

10.

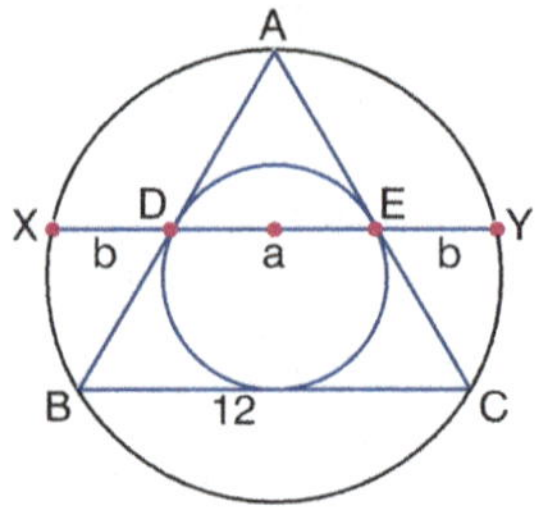

In the figure, ABC is an equilateral triangle and points D and E are tangent to the smaller circle.

Given that XD = EY = b and DE = a, find a/b.

11.

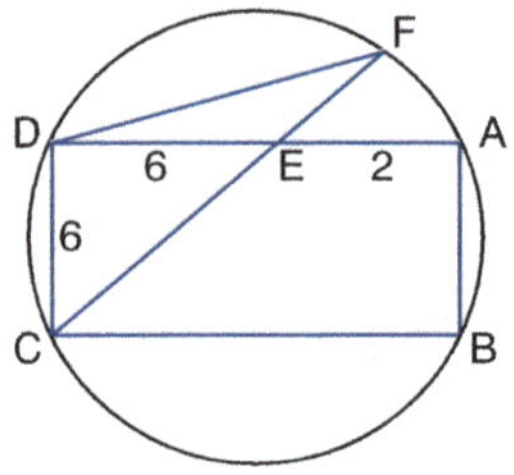

In the figure, ABCD is a rectangle.

CF intersects with DA at E.

If DC = DE = 6 and EA = 2, find BF.

12.

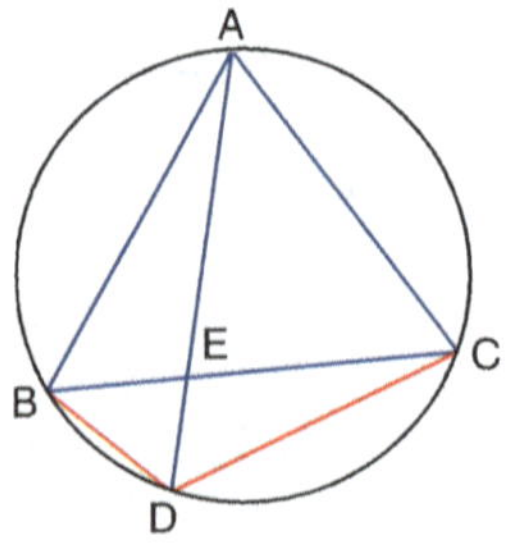

In the figure, ABC is an equilateral triangle.

If DC = 3BD, find AE/ED.

1. **(MathCON 2019 Final Gr12, #16)**

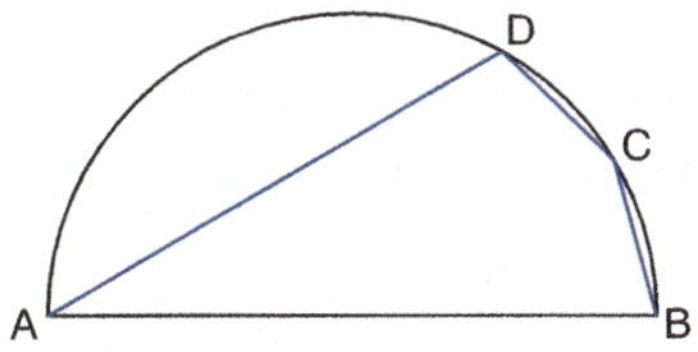

On a semicircle diameter $\overline{AB}$, two points C and D are taken such that BC = CD = 2.

If AB = 6, find the value of AD.

A) $\frac{16}{3}$ B) 5 C) $\frac{14}{3}$ D) 4 E) None of the preceding

2. **(MathCON 2021 Final Gr12, #39)**

Three circle of radii 6, 8 and r are externally tangent each of them to others. If the area of the triangle whose vertices are the centers of these circles is 84.

What is the perimeter of the triangle?

3. **(2012 AMC 10A, #18)**

Triangle ABC has AB = 27, AC = 26, and BC = 25.

Let I be the intersection of the internal angle bisectors of ΔABC.

What is BI?

A) 15 B) $5 + \sqrt{26} + 3\sqrt{3}$ C) $3\sqrt{26}$

D) $\frac{2}{3}\sqrt{546}$ E) $9\sqrt{3}$

4. **(2010 AMC 10B, #19)**

A circle center O has area 156π. Triangle ABC is equilateral, $\overline{BC}$ is a chord on the circle, OA = $4\sqrt{3}$, and point O is outside ΔABC.

What is the side length of ΔABC?

A) $2\sqrt{3}$ B) 6 C) $4\sqrt{3}$ D) 12 E) 18

5. **(2004 AMC 10B #24)**

In triangle ABC we have AB = 7, AC = 8, BC = 9.

Point ' is on the circumscribed of the triangle so that AD bisects angle BAC.

What is the value of $\frac{AD}{CD}$?

A) $\frac{9}{8}$ B) $\frac{5}{3}$ C) 2 D) $\frac{17}{7}$ E) $\frac{5}{2}$

6. **(2013 AMC 10A #23)**

In ΔABC, AB = 86, and AC = 97.

A circle with center A and radius AB intersects $\overline{BC}$ at points B and X. Moreover $\overline{BX}$ and $\overline{CX}$ have integer lengths.

What is BC?

A) 11 B) 28 C) 33 D) 61 E) 72

7. **(2016 AMC 10A #24)**

A quadrilateral is inscribed in a circle of radius $200\sqrt{2}$. Three of the sides of this quadrilateral have length 200.

What is the length of the fourth side?

A) 200 B) $200\sqrt{2}$ C) $200\sqrt{3}$ D) $300\sqrt{2}$ E) 500

8. **(2014 AMC 12B #24)**

Let ABCDE be a pentagon inscribed in a circle such that AB = CD = 3, BC = DE = 10, and AE = 14.

The sum of the lengths of all diagonals of ABCDE is equal to $\frac{m}{n}$, where m and n are relatively prime positive integers.

What is m + n?

A) 129 B) 247 C) 353 D) 391 E) 421

1. What is the radius?

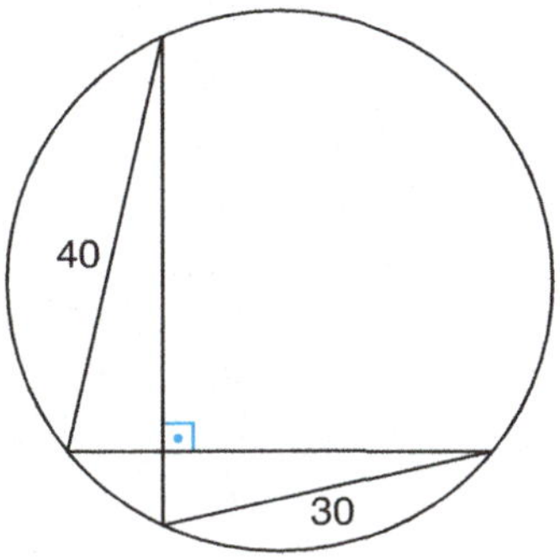

2.

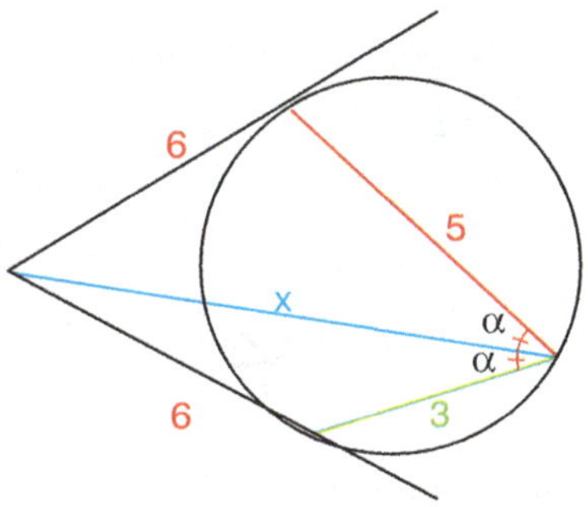

3.

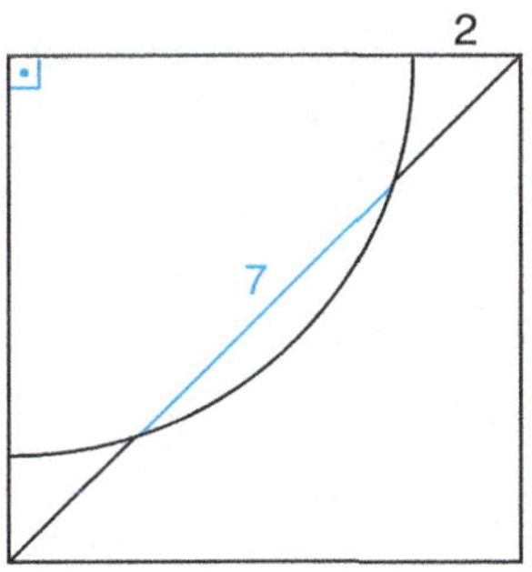

Radius of the square?

4.

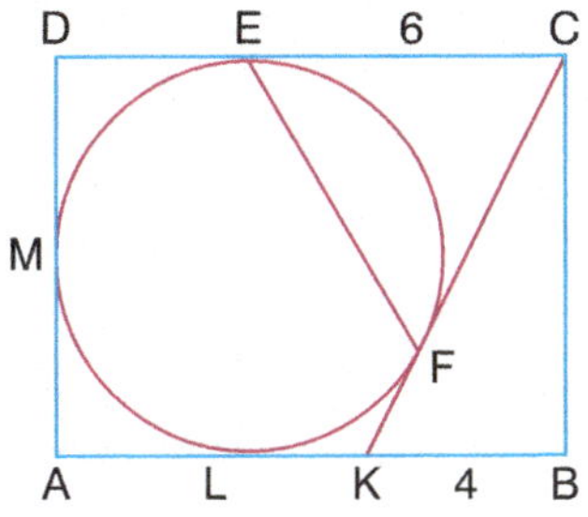

$\overline{EF} = ?$

5.

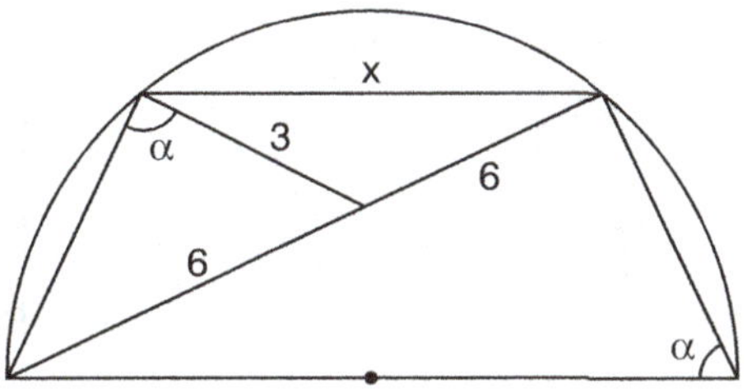

$x = ?$

6.

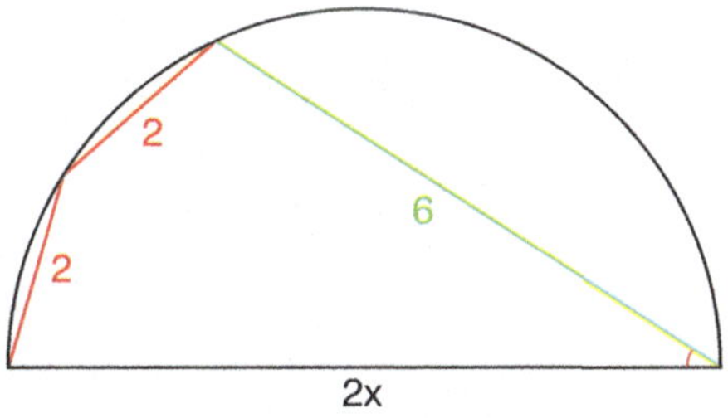

$x^2 - 3x = ?$

MathTopia Press

7.

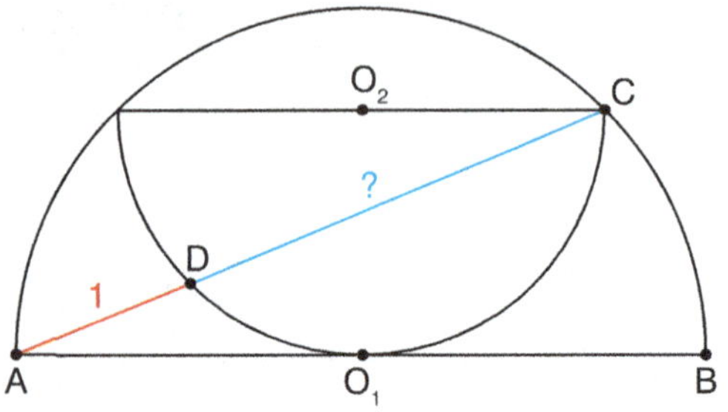

8.

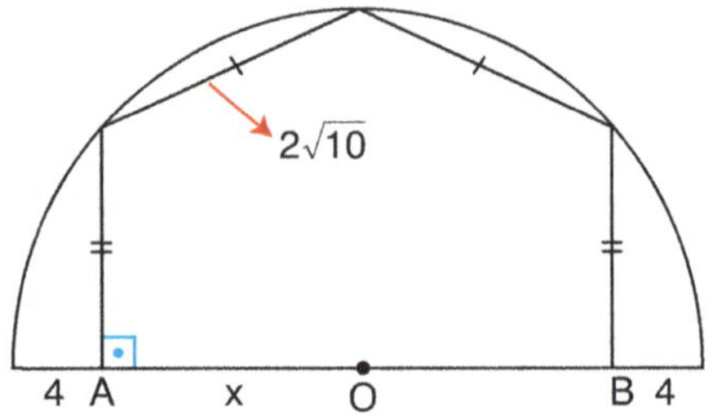

AB ?

9.

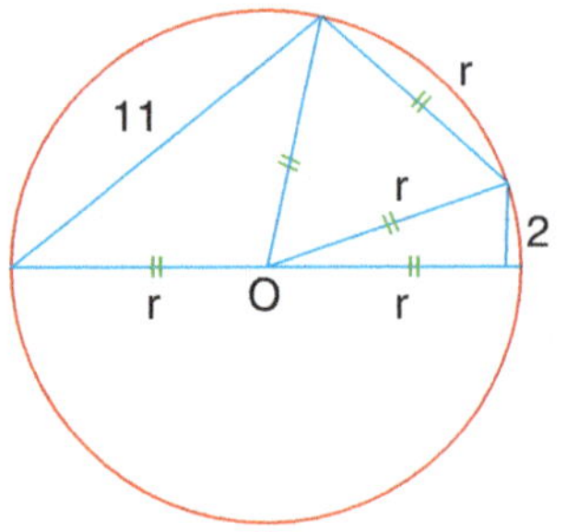

r = ?

10.

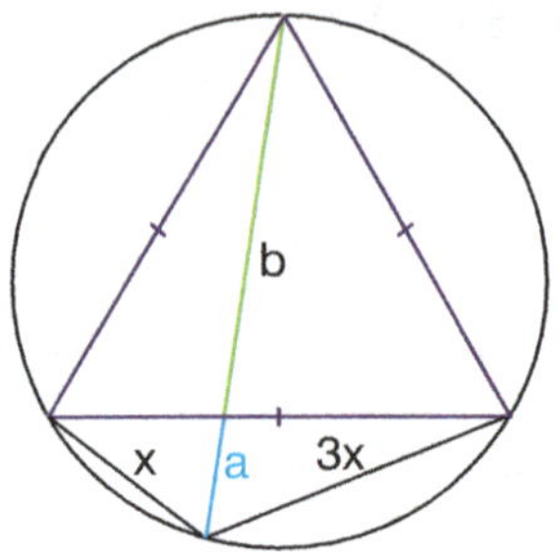

$\frac{a}{b} = ?$

11.

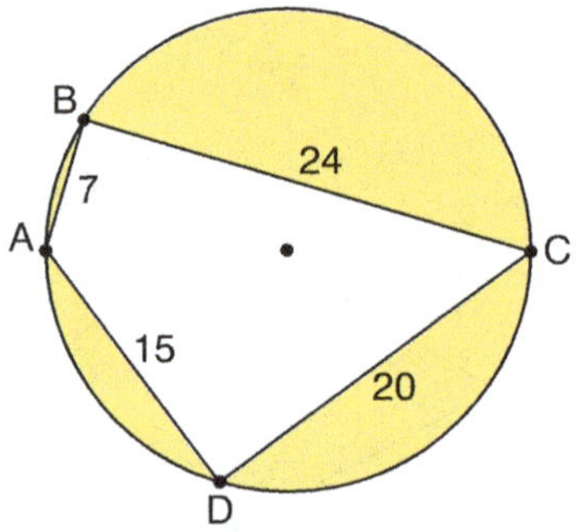

Yellow area = $(a\pi - b)/c$

a + b + c?

12.

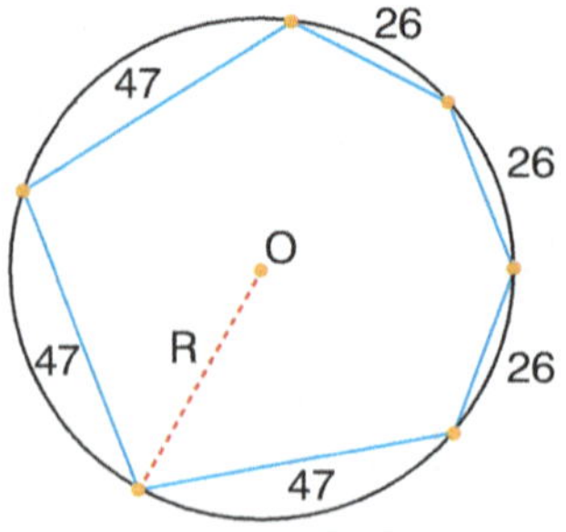

R = ?

CHAPTER 5

ANALYTIC GEOMETRY

Theorem 1.1

Distance between two points

The distance between two points $A(x_1, y_1)$ and $B(x_{2)}, y_{2)}$ is

$$AB = \sqrt{(x_2 - x_1)^2 + (y_2 - y_1)^2}$$

Proof 1.1

In the figure, ΔABC is a right triangle.

$AC = x_2 - x_1$

$BC = y_2 - y_1$

By the Pythagorean theorem,

$$AB^2 = AC^2 + BC^2$$

$$AB^2 = (x_2 - x_1)^2 + (y_2 - y_1)^2$$

and so $AB = \sqrt{(x_2 - x_1)^2 + (y_2 - y_1)^2}$ or

$$AB = \sqrt{(x_1 - x_2)^2 + (y_1 - y_2)^2}.$$

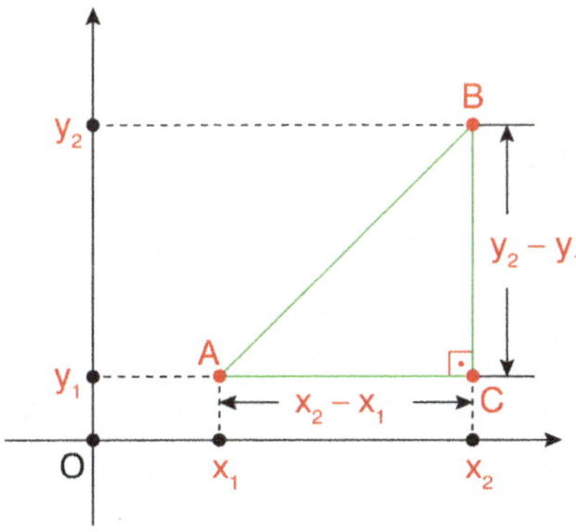

Theorem 1.2

Midpoint of a Line Segment

Let the points $A(x_1, y_1)$ and $B(x_2, y_2)$ be the endpoints of a line segment AB, and let $C(x_0, y_0)$ be the midpoint of AB. Then,

$$x_0 = \frac{x_1 + x_2}{2} \quad \text{and} \quad y_0 = \frac{y_1 + y_2}{2}.$$

Definition 1.1

Slope $m = \dfrac{y_2 - y_1}{x_2 - x_1}$ $(x_2 \neq x_1)$

Point – slope: $y - y_1 = m(x - x_1)$

Slope – intercept: $y = mx + b$

Standard: $Ax + By = C$ (with A, B not both 0)

Property 1.1

- Parallel lines: $m_1 = m_2$.
- Perpendicular lines: $m_1 m_2 = -1$ (non-vertical).

MathTopia Press

Example 1.1

Quadrilateral MATH has vertices with coordinates M(–1, 7), A(3, 5), T(2, –7), and H(–6, –3). Prove that quadrilateral MATH is a trapezoid.

$$m_{MA} = \frac{5 - 7}{3 - (-1)} = \frac{-2}{4} = -\frac{1}{2},$$

$$m_{TH} = \frac{-3 - (-7)}{-6 - 2} = \frac{4}{-8} = -\frac{1}{2},$$

so $\overline{MA} \parallel \overline{TH}$.

For the other pair,

$$m_{AT} = \frac{-7 - 5}{2 - 3} = 12, \qquad m_{MH} = \frac{-3 - 7}{-6 - (-1)} = 2,$$

which are unequal; hence $\overline{AT} \parallel \overline{MH}$.

Thus **exactly one** pair of opposite sides is parallel, so MATH is a **trapezoid**.

Theorem 1.3

Angle between lines: $\tan\theta = \left|\dfrac{m_2 - m_1}{1 + m_1 m_2}\right|.$

Theorem 1.4

Theorem – Distance from a Point to a Line

Let L be the line

$$ax + by + c = 0$$

and let $Q(x_0, y_0)$ be a point in the plane.

Then the perpendicular distance d from Q to L is given by

$$d = \frac{|ax_0 + by_0 + c|}{\sqrt{a^2 + b^2}}.$$

Proof 1.4

The line $ax + by + c = 0$ can be rewritten as

$$y = -\frac{a}{b}x - \frac{C}{b},\ (b \neq 0),$$

so its slope is $-\dfrac{a}{b}$.

A line perpendicular to L has slope $\dfrac{b}{a}$ (since $-\dfrac{a}{b} \cdot \dfrac{b}{a} = -1$).

Let the equation of the perpendicular line passing through $Q(x_0, y_0)$ be

$$y - y_0 = \frac{b}{a}(x - x_6).$$

To find the foot of the perpendicular $P(x_1, y_1)$, solve the system

$$\begin{cases} ax_1 + by_1 + c = 0, \\ y_1 - y_0 = \dfrac{b}{a}(x_1 - x_0). \end{cases}$$

Multiplying the second equation by a and substituting y_1 into the first, we get

$$ax_1 + b\left(y_0 + \frac{b}{a}(x_1 - x_0)\right) + c = 0,$$

$$x_1(a^2 + b^2) = a(by_0 - c) + b^2x_0.$$

Proof 1.4

Hence

$$x_1 = \frac{b^2x_0 - aby_0 - ac}{a^2 + b^2},\quad y_1 = \frac{-abx_0 + a^2y_0 - bc}{a^2 + b^2}.$$

The vector $\overrightarrow{OP} = (x_1 - x_0, y_1 - y_0)$ represents the perpendicular from Q to L.

$$d = \sqrt{(x_1 - x_0)^2 + (y_1 - y_0)^2}.$$

Substituting x1, y1 and simplifying gives

$$d = \frac{|ax_0 + by_0 + c|}{\sqrt{a^2 + b^2}}.$$

Thus, the perpendicular distance from $Q(x_0, y_0)$ to $ax + by + c = 0$ is as stated.

$$d = \frac{|ax_0 + by_0 + c|}{\sqrt{a^2 + b^2}}.$$

Example 1.2

Distance from (3, 1) to $2x + y = 17$.

Write $2x + y - 17 = 0$.

$$d = \frac{|2 \cdot 3 + 1 - 17|}{\sqrt{2^2 + 1^2}} = \frac{10}{\sqrt{5}} = 2\sqrt{5}.$$

Example 1.3

Let $C(x_0, y_0)$ divide AB internally in the ratio

$$\frac{CA}{CB} = 3.$$

Then CA = 3a and CB = a, so the line is divided into **4 equal parts**.

Solution:

To find x_0:

Change in x from A(–3, 5) to B(9, –11) is 12.

Since AC = 3/4 of AB:

$$x_0 = -3 + \frac{3}{4}(12) = -3 + 9 = 6.$$

To find y_0:

Change in y from A to B is – 16.

So, $y_0 = 5 + \dfrac{3}{4}(-16) = 5 - 12 = -7.$

Example 1.4

Let $P(x_p, y_p)$ divide MN **externally** in the ratio

$$\frac{PM}{PN} = \frac{3}{2}.$$

Solution:

To find x_p:

The change in x from M(–2, 3) to N(4, 5) is + 6.

Since P divides externally, extend the line by $\frac{3}{2}$ of MN:

$$x_p = -2 + 3 \times 6 = 16.$$

To find y_p:

The change in y from M to N is + 2.

Extend by the same external ratio:

$$y_p = 3 + 3 \times 2 = 9.$$

Theorem 1.5

Distance between two parallel lines

Let d_1: $ax + by + c_1 = 0$ and d_2: $ax + by + c_2 = 0$ be two parallel lines. Then the distance between d_1 and d_2 is

$$l = \frac{|c_2 - c_1|}{\sqrt{a^2 + b^2}}$$

Proof 1.5

The distance of any point A(x, y) on line d_1 to line d_2 is

$$\frac{|ax + by + c_2|}{\sqrt{a^2 + b^2}}.$$

In the equation of d_1: $ax + by + c = 0$

$ax + by \; 0 \; -c_1$, and so $l = \frac{|c_2 - c_1|}{\sqrt{a^2 + b^2}}$

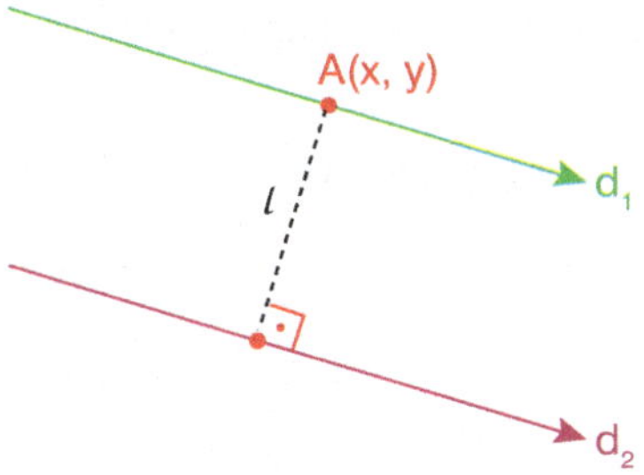

Example 1.5

Find the distance between the parallel lines $x - 2y + 5 = 0$ and $3x - 6y + 9 = 0$.

Solution:

$$\left.\begin{array}{l} d_1: x - 2y + 5 = 0 \\ d_2: 3x - 6y + 9 = 0 \end{array}\right\} \left.\begin{array}{l} 3x - 6y + 15 = 0 \\ 3x - 6y + 9 = 0 \end{array}\right\} \begin{array}{l} d_1 \text{ was} \\ \text{multiplied} \\ \text{by 3.} \end{array}$$

Now, we have

$$\left.\begin{array}{l} c_1: 15 \\ c_2: 9 \end{array}\right\} l = \frac{|c_1 - c_2|}{\sqrt{a^2 + b^2}} = \frac{15 - 9}{\sqrt{9 + 36}} = \frac{6}{\sqrt{45}} = \frac{6}{3\sqrt{5}} = \frac{2\sqrt{5}}{5}.$$

Example 1.6

Find the ordinate of the point on the y-axis which is equidistant to the points A(–4, 0) and B(9, 5).

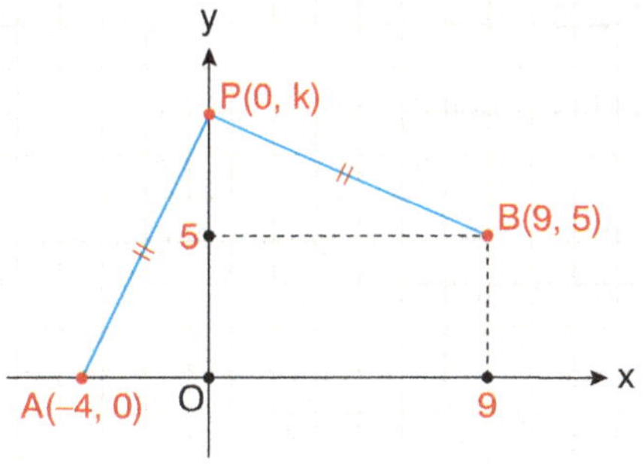

Solution:

The point is on the y-axis, so its x-coordinate is 0. Let us call the point P(0, k). Now, from the diagram, PA = PB

$$\sqrt{(-4)^2 + k^2} = \sqrt{9^2 + (k - 5)^2}$$

$$(-4)^2 + k^2 = 9^2 + (k - 5)^2$$

$$16 + k^2 = 81 + k^2 - 10k + 25$$

$$10k = 90$$

$k = 9$. Therefore, the point is P(0, 9).

Theorem 1.6

Standard equation of a circle

In the coordinate plane, the equation of a circle with center C(a, b) and radius r is $(x - a)^2 + (y - b)^2 = r^2$.

Proof 1.6

Let P(x, y) be any point on the circle,

C(a, b) be the center and r be the radius of the circle.

Since r = CP, by using the distance formula we can write $CP = \sqrt{(x-a)^2 + (y-b)^2} = r$.

By taking the square of both sides of the equation, we get

$$(x-a)^2 + (y-b)^2 = r^2.$$

This is the standard equation of a circle.

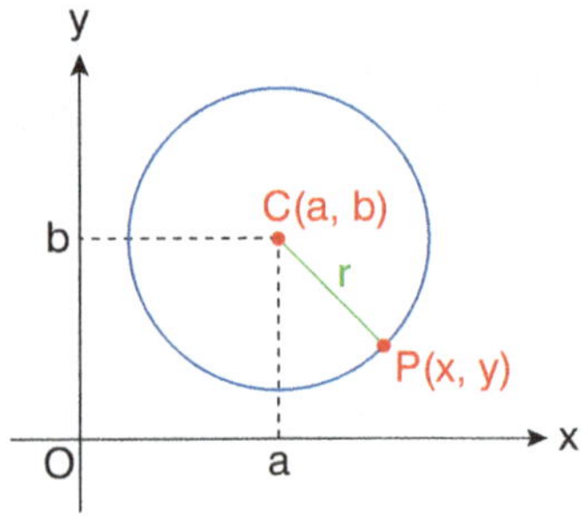

Definition 1.2

(a) Translation

A translation by the vector (a, b) moves any point (x, y) to

$$(x', y') = (x + a, y + b).$$

(b) Rotation

A rotation by an angle α (counterclockwise) about a fixed point P moves a point A to a point B such that

$$\angle APB = \alpha \text{ and } PA = PB.$$

(c) Reflection about a Point

A reflection about a point P sends A to B so that P is the **midpoint** of segment AB.

(d) Reflection about a Line

A reflection about a line l sends A to B such that l is the perpendicular bisector of segment AB.

Example 1.7

Show that the equation

$$2x^2 + 2y^2 - 8x + 6y + 8 = 0$$

represents a circle, and find its **center** and **radius** by completing the square.

Solution:

Divide through by 2 to simplify:

$$x^2 + y^2 - 4x + 3y + 4 = 0.$$

Group and complete the squares:

$$(x^2 - 4x) + (y^2 + 3y) = -4.$$

$$(x-2)^2 - 4 + (y + \frac{3}{2})^2 - \frac{9}{4} = -4.$$

Simplify:

$$(x-2)^2 + (y + \frac{3}{2})^2 = \frac{9}{4}.$$

My Notes:

Example 1.8

A point P(a, b) is rotated 90° counterclockwise about (1, 5) and then reflected in the line y = –x. The final image is (–6, 3). Find a, b, and b – a.

Solution:

Translate to the center: (a – 1, b – 5).

Rotate 90° CCW: (–(b – 5), a – 1).

Translate back: (6 – b, a + 4).

Reflect in y = –x: (– (a + 4), –6, b).

Set equal to (–6, 3):

$$-(a + 4) = -6 \Rightarrow a = 2, \qquad -6 + b\ )\ 3 \Rightarrow b = 9.$$

Thus b – a = 9 – 2 = 7 (with a = 2, b = 9).

Theorem 1.7

Theorem – Shoelace Formula for the Area of a Polygon

Let a polygon in the coordinate plane have consecutive vertices

$$(x_1, y_1), (x_2, y_2), (x_3, y_3), ..., (x_n, y_n).$$

Then the area A of the polygon is given by

$$A = \frac{1}{2}|(x_1y_2 + x_2y_3 + ... + x_{n-1} \cdot y_n + x_ny_1) -$$
$$(y_1x_2 + y_2x_3 + ... + y_n - 1x_n + y_nx_1)|.$$

Example 1.9

Find the area of the triangle with vertices

A(1, 3), B(4, 12), C(9, 16).

Solution:

Using the Shoelace Formula,

$A=\frac{1}{2}|(1 \times 12 + 4 \times 16 + 9 \times 3)-(3 \times 4 + 12 \times 9 + 16 \times 1)|.$

$A=\frac{1}{2}|(12 + 64 + 27)-(12 + 108+ 16)|=\frac{1}{2}|103-136|=\frac{33}{2}.$

Practice 1

Find the area of ΔTRY.

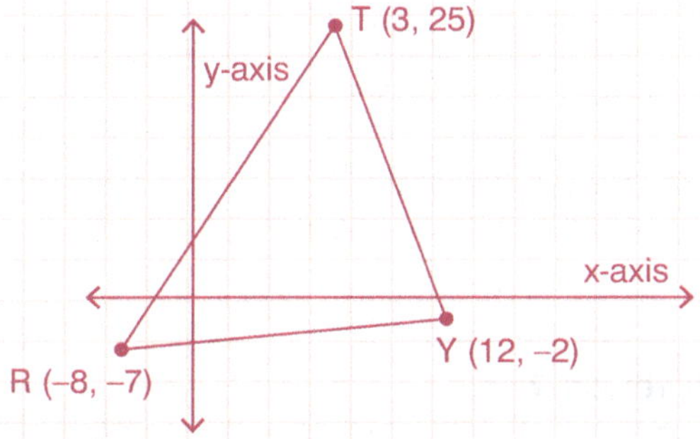

Solution:

Method 1 – Shoelace

$A = \frac{1}{2}|(x_T y_Y + x_Y y_R + x_R y_T - (y_T x_Y + y_Y x_R + y_R x_T)|$

$=\frac{1}{2}(3)(-2)+(12)(-7)+(-8)(25)-[(25)(12)+(-2)(-8)+(-7)(3)]|$

$=\frac{1}{2}|-6-84-200-300(300+16-21)|=\frac{1}{2}|-290-295|=\frac{585}{2}.$

Method 2 – Base–height (nice cancellation)

Line Y R: through (12, –2) and (–8, –7)

$\Rightarrow 5x - 20y - 100 = 0.$

Base Y R =

Height from T to Y R:

$$h = \frac{|5 \cdot 3 - 20 \cdot 25 - 100|}{\sqrt{5^2 + (-20)^2}} = \frac{585}{5\sqrt{17}}.$$

Thus $A = \frac{1}{2} \cdot 5\sqrt{17} \cdot \frac{585}{5\sqrt{17}} = \frac{585}{2} = 295.5.$

MathTopia Press

Practice 2

Triangle A(–2, 6), B(4, 17), C(x, 11) has area 42. Find all x.

Solution:

$\text{Area} = \frac{1}{2}|(x_1(y_2 - y_3) + x_2(y_3 - y_1) + x_3(y_1 - y_2)|$

$42 = \frac{1}{2}|(-2)(17 - 11) + 4(11 - 6) + x(6 - 17)|$

$= \frac{1}{2}|-12 + 20 - 11x| = \frac{1}{2}|8 - 11x|$

$|8 - 11x| = 84 \Rightarrow 8 - 11x = \pm 84.$

Thus $x = \frac{-76}{11}$ or $x = \frac{92}{11}$.

Practice 3

In the xy-plane, the set of all points (x, y) that satisfy

$$\sqrt{(x + 3)^2 + (y - 2)^2} = \sqrt{(x - 3)^2 + y^2}$$

represents what type of curve?

Solution:

- LHS is distance from (–3, 2). RHS is distance from (3, 0).
- Equation means points equidistant from those two points $\Rightarrow$ **perpendicular bisector** of segment.
- Squaring and simplifying:

  $(x + 3)^2 + (y - 2)^2 = (x - 3)^2 + y^2$

  $\Rightarrow 3x - y + 1 = 0.$
- Final answer: **a line**.

Practice 4

A circle has two parallel tangents at $P_1 = (2, 5)$ and $P_2 = (8, 13)$. Find their common slope.

Solution:

If the tangent slope is m, then each radius to the tangent point has slope $-\frac{1}{m}$.

The center lies on lines through P_1, P_2 with slope $-\frac{1}{m}$; hence the line $P_1 P_2$ it self has slope $-\frac{1}{m}$:

$\text{slope}(P_1 P_2) = \frac{13-5}{8-2} = \frac{8}{6} = \frac{4}{3} = -\frac{1}{m} \Rightarrow m = -\frac{3}{4}.$

So the common slope is $-\frac{3}{4}$.

Practice 5

In the square OABC with side length 2a, points D and E are located on sides OC and BC, respectively, such that

$$|OD| = |EC|.$$

The diagonals AE and BD intersect point F.

Find the relationship between the coordinates (x, y) of point F.

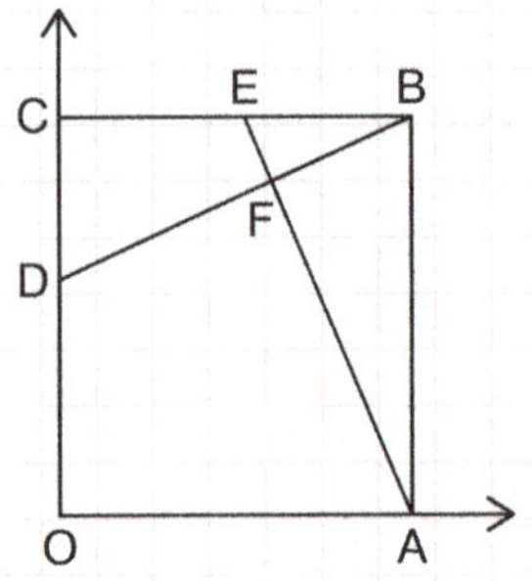

Solution:

Let

$$O(0, 0),\quad A(2a, 0),\quad B(2a, 2a),\quad C(0, 2a).$$

Then

$$D(0, t),\quad E(t, 2a),\quad \text{where } 0 < t < 2a.$$

Equations of lines

$$AE: y = \frac{2a}{t-2a}(x-2a),\quad BD: y-2a = \frac{t-2a}{-2a}(x-2a).$$

Solving for intersection F(x, y) and eliminating t gives

$$(x-2a)^2 + (y-a)^2 = a^2.$$

My Notes:

Practice 6

The equation of d in the figure is $x + y - 6 = 0$.

OABC is a square and OA = 2 units. Given $BH \perp KL$, what is BH?

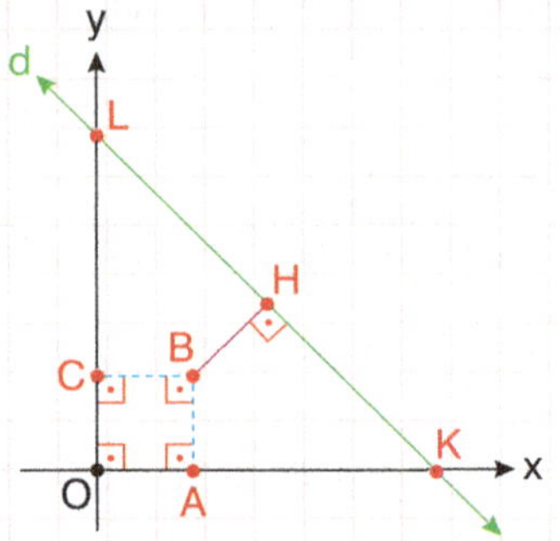

Work Space

Practice 7

Lines $y = 4x$ and $y = \frac{1}{2}x$ intersect at the origin.

A **square** ABCD is formed so that its **diagonals** lie on these two lines.

Find the equations of its diagonals and coordinates of its vertices in simplest from

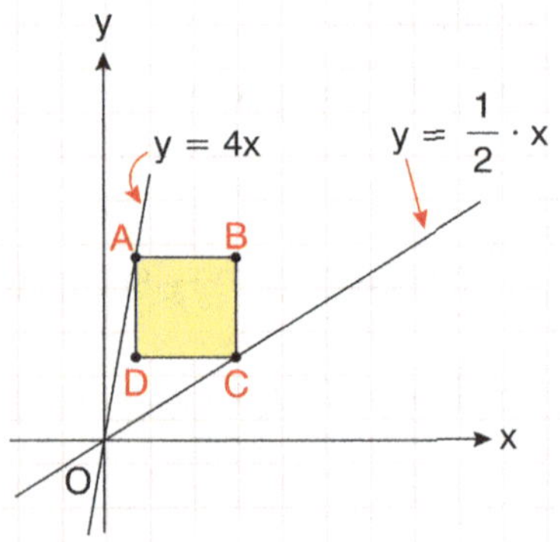

Work Space

1. Rectangle ABCD has two vertices at coordinates A(–1, –3) and B(6, 5). The slope of $\overline{BC}$ is

(1) $-\frac{7}{8}$ (3) $-\frac{8}{7}$

(2) $\frac{7}{8}$ (4) $\frac{8}{7}$

2. In the coordinate plane, the reflection of point (4, 4) across a line passing through (1, 0) is (a, 0). What is the product of all possible values of a?

3. Given that a ≠ 0. Lines $ax + (b + 1)y - 1 = 0$ and $(b - 5)x + ay + 7 = 0$ are perpendicular to each other. Find b.

4. Given that a can be neither 2 nor 1.

Lines $(a + 2)x + 2y + 4 = 0$ and $2x + (a - 1)y + 6 = 0$ are parallel to each other. Find the possible negative value of a.

5. In the figure below, B = (a, b) is the intersection point of two lines. Find a + b.

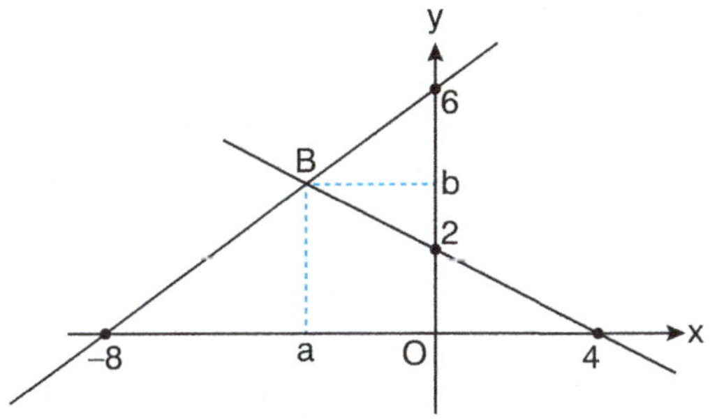

6.

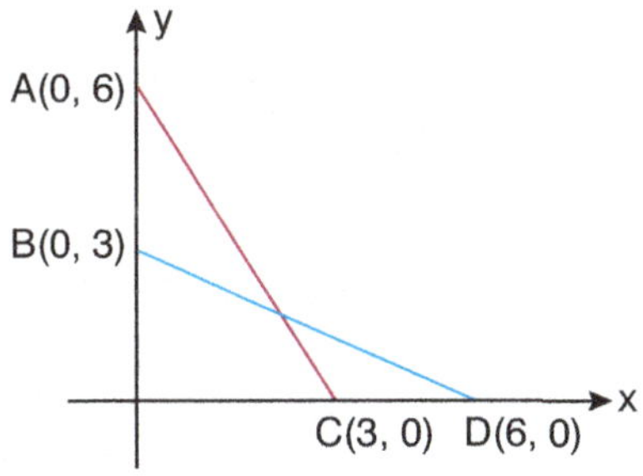

In the coordinate plane given above, which of the following is the coordinate of the intersection point K of the line segment [AC] and [BD] ?

A) (2, 1) B) $(2, \frac{3}{2})$ C) (2, 2) D) $(1, \frac{1}{2})$ E) (1, 2)

7. Use the diagram below to answer the question that follows.

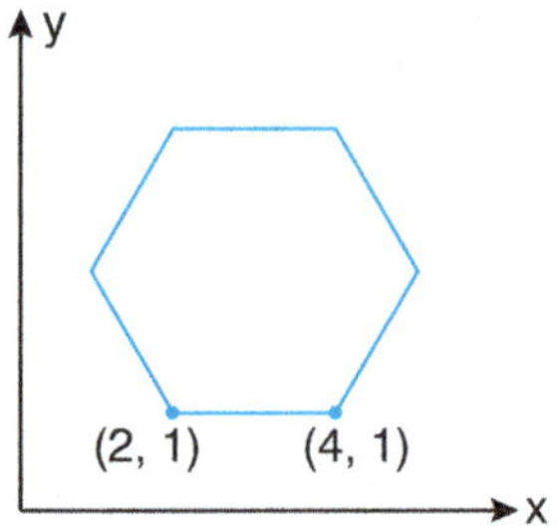

A regular hexagon has two of its vertices at the points (2,1) and (4, 1) on a coordinate grid, as shown in the diagram above. What is the area of the hexagon?

8.

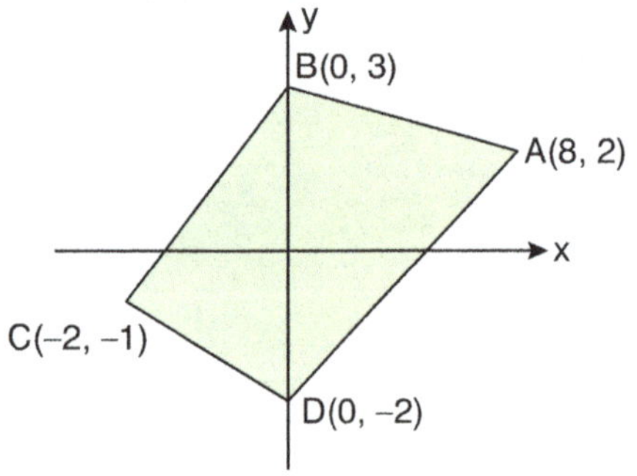

What is the area of the quadrilateral ABCD = ?

9. The vertices of a triangle are (–2, 6), (4, 17), and (x, 11). The area of the triangle is 42. Find all possible values of x.

10. (AMC 10 2006 A, #8)

A parabola with equation.

$$y = x^2 + bx + c$$

passes through the points (2, 3) and (4, 3). What is the value of c?

11. (BmMT 2012 Cindering, #14)

What is the minimum distance from the point (1, 2) to the line

$$3x + 4y = 5?$$

12. What is the length of the longest chord of the circle

$$x^2 + 2x + y^2 - 6y = 6.$$

1. In the rectangular coordinate plane, the line $y = mx$ intersects the circle

$$x^2 - 26x + y^2 + 144 = 0$$

at two distinct points.

Accordingly, which of the following intervals shows all possible values of m?

A) $(-\frac{3}{4}, \frac{3}{4})$ B) $(-\frac{3}{8}, \frac{3}{8})$ C) $(-\frac{4}{9}, \frac{4}{9})$

D) $(-\frac{5}{12}, \frac{5}{12})$ E) $(-\frac{7}{24}, \frac{7}{24})$

2. A line is tangent to a circle at the point (2, 5). A parallel line is tangent to the same circle at the point (8, 13). What is the common slope of the two tangent lines? Express your answer as a common fraction.

3. **(2026 AMC 10B, #6)**

In the coordinate plane, consider the square region

$$0 \le x \le 2,\ 0 \le y \le 2.$$

The line

$$y = \frac{1}{3}x + 1$$

splits this square into an upper region and a lower region. A vertical line $x = a$ is drawn so that it divides the lower region into two regions of equal area.

If a can be written in the form $\sqrt{s} - t$, where s and t are positive integers, find $s + t$.

4. In the coordinate plane, one vertex of a triangle is at the origin. The other two vertices lie on the lines $y = x$ and $y = -x$. The medians of the triangle intersect at the point (2, 4). What is the area of the triangle?

5. An equation of circle M is $x^2 + y^2 + 6x - 2y + 1 = 0$. What are the coordinates of the center and the length of the radius of circle M?

A) center, (3, –1) and radius 9

B) center, (3, –1) and radius 3

C) center, (–3, 1) and radius 9

D) center, (–3, 1) and radius 3

6.

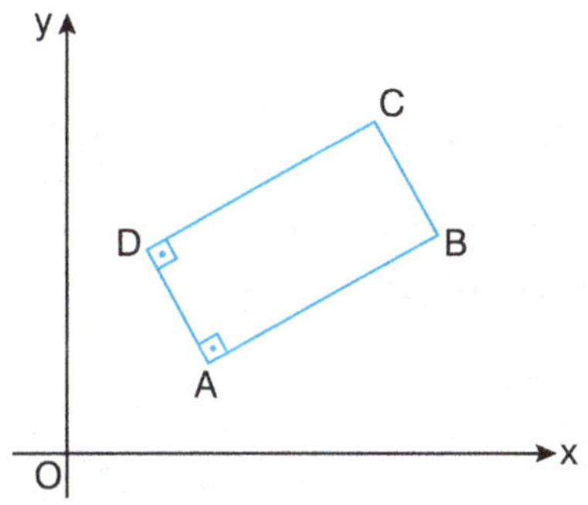

ABCD is a rectangle.

The equation of the straight line DA is $2x + y = 8$

The coordinates of C are (5, 10)

Find the equation of the straight line CD.

7. **(2004 National Mathcounts Sprint Round)**

The intersection of the graphic solutions of all three inequalities listed below form an enclosed region in the coordinate plane. What is the area of this region ? Express your answer in terms of π.

$$(x-4)^2 + y^2 \leq 16$$
$$y \geq x - 4$$
$$y \geq -\frac{1}{3}x$$

8. Let n be a positive real number. Consider the lines

$$d_1 : y = x + 4 \text{ and } d_2 : y = 3x + n.$$

Together with the coordinate axes, these two lines form two bounded triangular regions: one between the lines and the x-axis, and the other between the lines and the y-axis. If the areas of these two triangles are equal, what is the value of n?

9. In the coordinate plane, the line $y = x$ intersects square ABCD at points K(2, 2) and L(7, 7). Vertex B lies on the y-axis, and vertex C lies on the x-axis. What are the coordinates of vertex D?

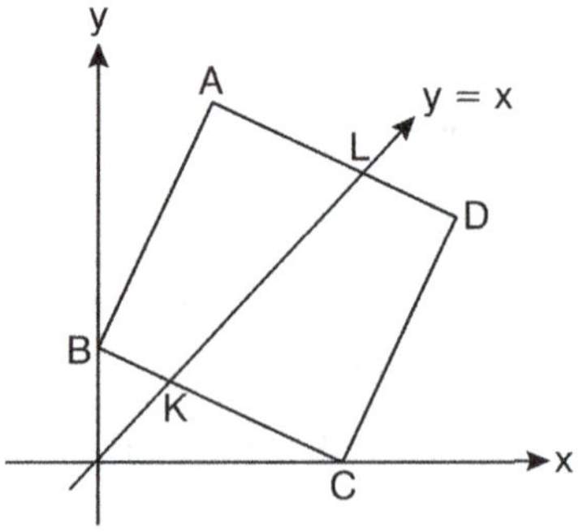

10. Line d_1 passes through (4, 0) and (0, –3). Line d_2 passes through (–2, 0) and (0, b), where b is a positive real number. The two lines intersect at point E.

The triangles with vertices (0, 0), (0, b), (–2, 0) and (0, 0), (4, 0), (0, –3) have equal areas. What is the sum of the coordinates of point E?

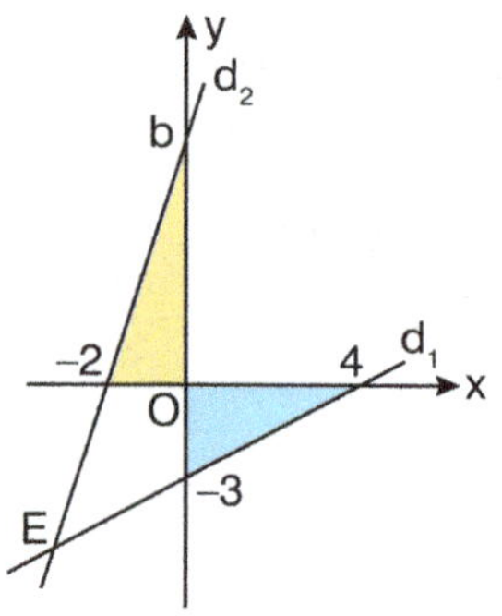

11. In the coordinate plane, triangle ABC is an isosceles right triangle with right angle at A. Point C has coordinates (6, 12). The midpoint of $\overline{AB}$ lies on the y-axis. What is the sum of the coordinates of point A?

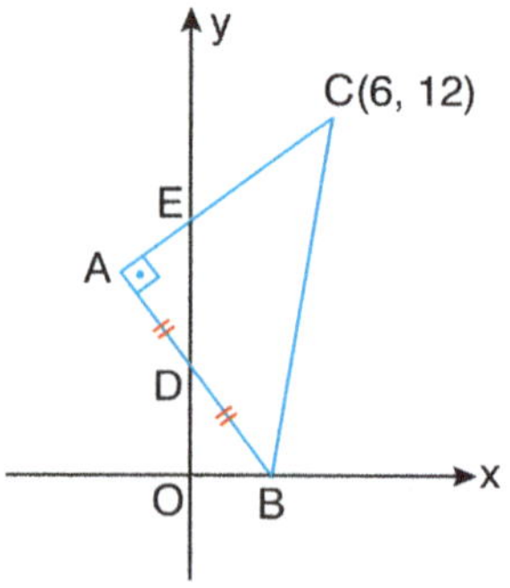

12.

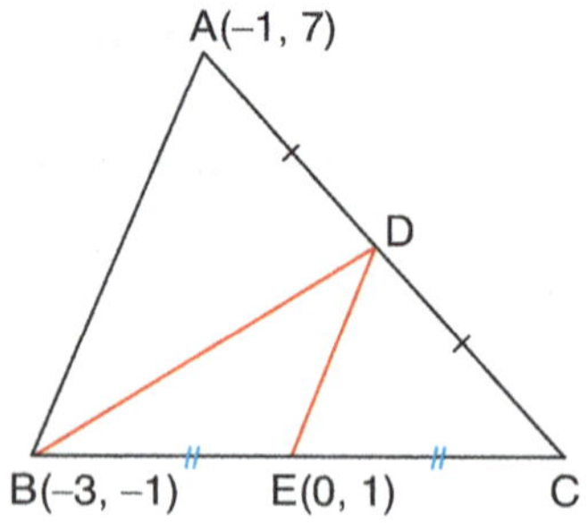

In triangle ABC

$|AD| = |CD|$ and $|BE| = |CE|$

Given the above, how many units is $|BD|$?

1. (MathCON 2020 Online GR9, #16)

In the diagram, $\overline{YX} \perp \overline{YZ}$.

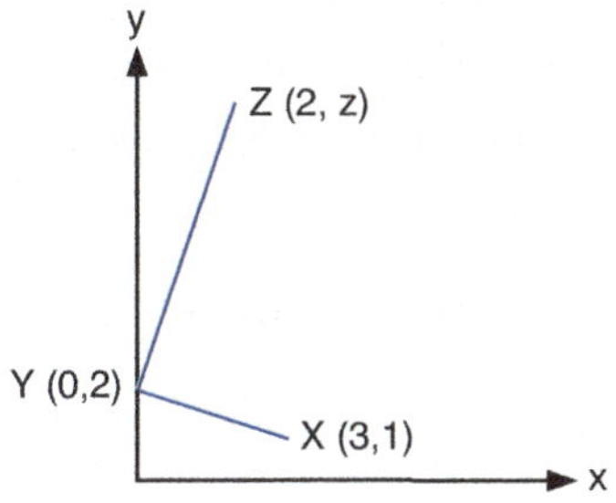

What is the value of z?

A) 5 B) 6 C) 7 D) 8 E) 9

2. (2021 AMC 12B, #5)

The point P(a, b) in the xy plane is first rotated counterclockwise by 90° around the point (1, 5) and then reflected about the line $y = -x$. The image of P after these two transformations is at (–6, 3).

What is b – a?

A) 1 B) 3 C) 5 D) 7 E) 9

3. (2016 AMC 12A, #7)

Which of these describes the graph of

$x^2(x + y + 1) = y^2(x + y + 1)$?

A) two parallel lines

B) two intersecting lines

C) three lines that all pass through a common point

D) three lines that do not pass through a common point

E) a line and a parabola

4. (2017 AMC 10B, #10)

The lines with equations $ax - 2y = c$ and $2x + by = -c$ are perpendicular and intersect at (1, 5).

What is C?

A) –13 B) –8 C) 2 D) 8 E) 13

5. (2019 AMC 10A, #7)

Two lines with slopes $\frac{1}{2}$ and 2 intersect at (2, 2).

What is the area of the triangle enclosed by these two lines and the line $x + y = 10$?

A) 4 B) $4\sqrt{2}$ C) 6 D) 8 E) $6\sqrt{2}$

6. (2013 AMC 12B, #8)

Line l_1 has equation $3x - 2y = 1$ and goes through $A = (-1, -2)$.

Line l_2 has equation $y - 1$ and meets line l_1 at point B.

Line l_3 has positive slope, goes through point A, and meets l_2 at point C. The area of ABC is 3.

What is the slope of l_3?

A) $\frac{2}{3}$ B) $\frac{3}{4}$ C) 1 D) $\frac{4}{3}$ E) $\frac{3}{2}$

7. (2023 AMC 10B, #13)

What is the area of the region in the coordinate plane defined by $||x| - 1| + ||y| - 1| \le 1$?

A) 2 B) 8 C) 4 D) 15 E) 12

8. (2009 AMC 12B, #9)

Triangle ABC has vertices A = (3, 0), B = (0,3), and C, where C is on the line x + y = 7.

What is the area of ΔABC?

9. (2013 AMC 12A, #13)

Let points A = (0, 0), B = (1, 2), C = (3, 3), and D = (4, 0).

Quadrilateral ABCD is cut into equal area pieces by a line passing through A. This line intersects $\overline{BC}$ at point $\left(\frac{p}{q}, \frac{r}{s}\right)$, where these fractions are in lowest terms.

What is p + q + r + s?

A) 54 B) 58 C) 62 D) 70 E) 75

10. (2007 AMC 12A, #13)

A piece of cheese is located at (12, 10) in a coordinate plane. A mouse is at (4, –2) and is running up the line y = –5x + 18. At the point (a, b) the mouse starts getting farther from the cheese rather than closer to it.

What is a + b?

A) 6 B) 10 C) 14 D) 18 E) 22

11. (2012 AMC 12B, #17)

Square PQRS lies in the first quadrant. Points (3, 0), (5, 0), (7, 0), and (13, 0) lie on lines SP, RQ, PQ, and SR, respectively.

What is the sum of the coordinates of the center of the square PQRS?

A) 6 B) $\frac{31}{5}$ C) $\frac{32}{5}$ D) $\frac{33}{5}$ E) $\frac{34}{5}$

12. (2015 AMC, #17)

A line that passes through the origin intersects both the line

x = 1 and the line $y = 1 + \frac{\sqrt{3}}{3}x$.

The three lines create an equilateral triangle.

What is the perimeter of the triangle?

A) $2\sqrt{6}$ B) $2 + 2\sqrt{3}$ C) 6 D) $3 + 2\sqrt{3}$ E) $6 + \frac{\sqrt{3}}{3}$

1.

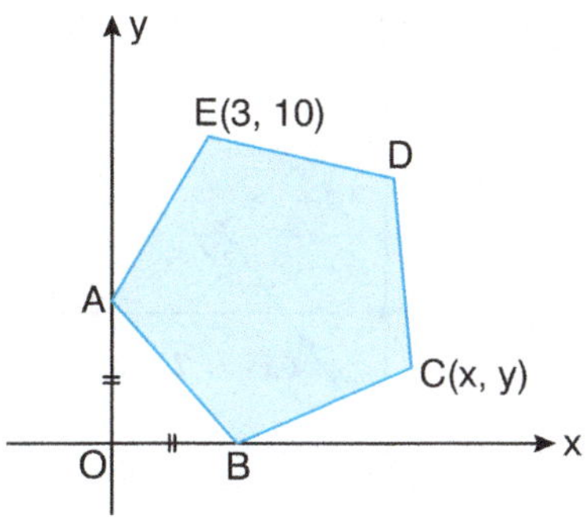

$x + y = ?$

2.

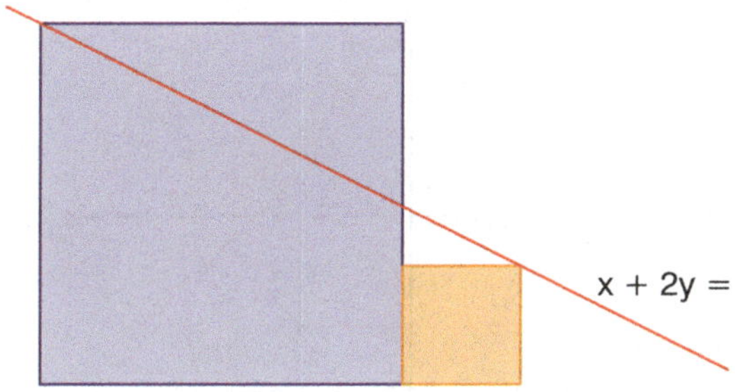

Area of the smaller square / area of the larger square = ?

3.

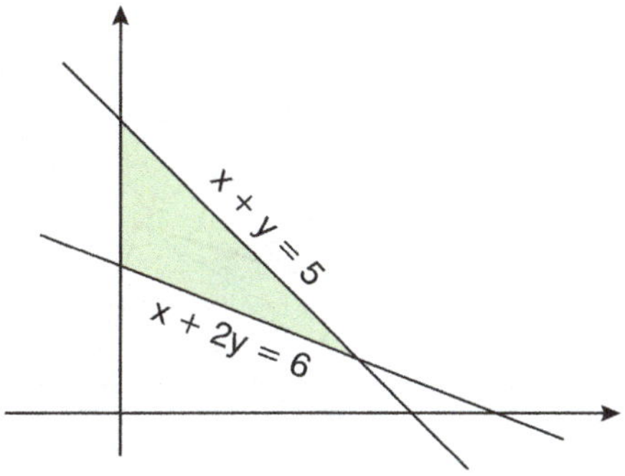

Shaded area = ?

4.

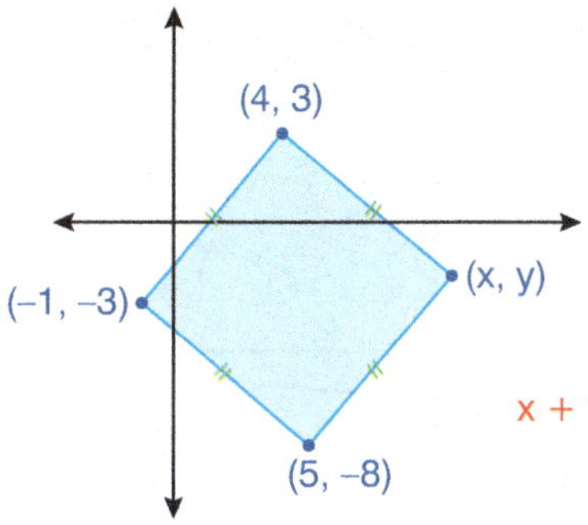

$x + y = ?$

5.

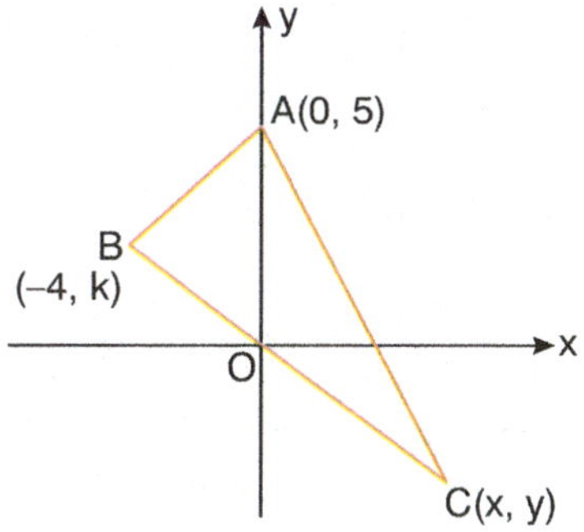

$OA = OB = OC$

$x + y = ?$

6.

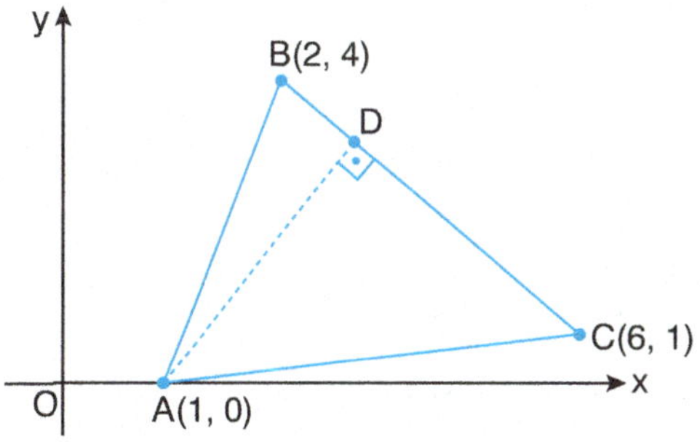

Find AD

7.

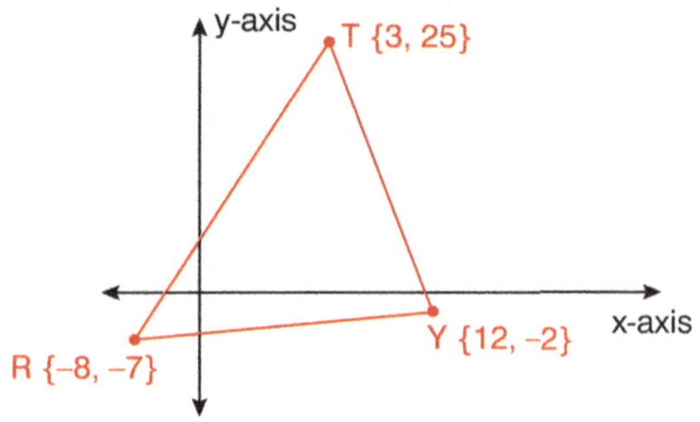

Area of $\triangle TRY$ = ?

8.

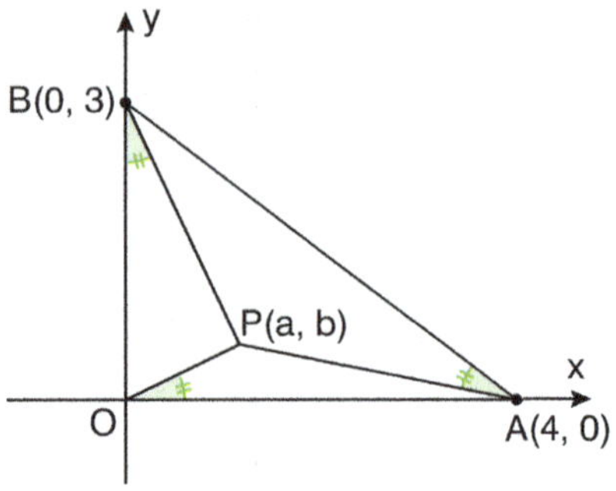

a = ?

9. Distance from (3,1) to the line 2x + y = 17

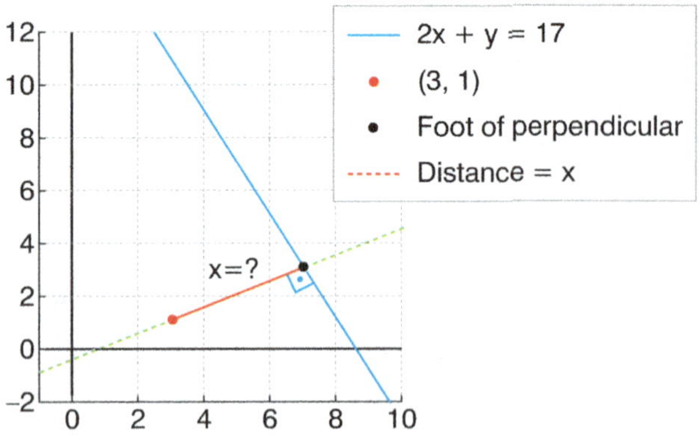

10.

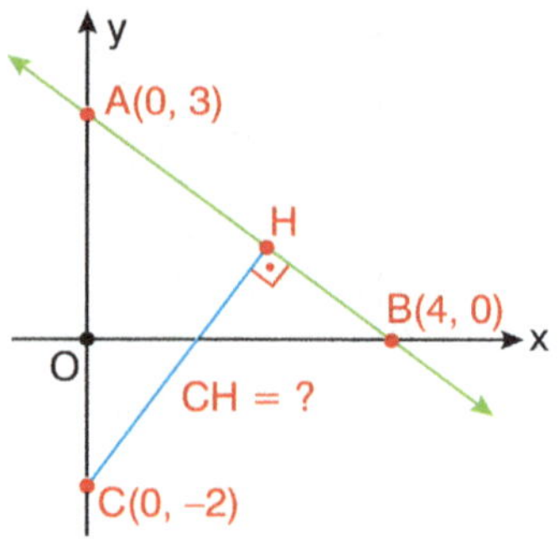

11.

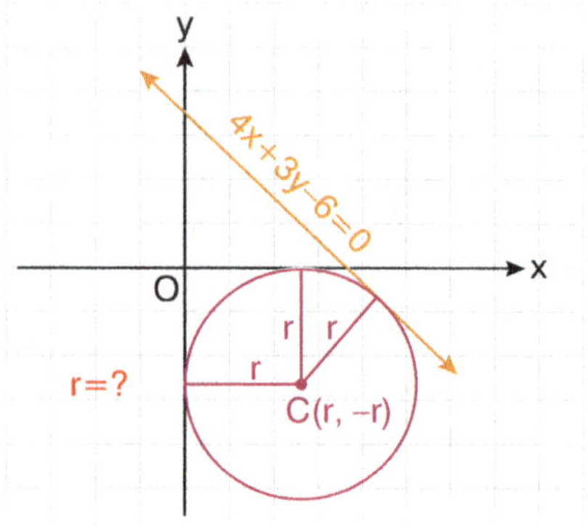

12.

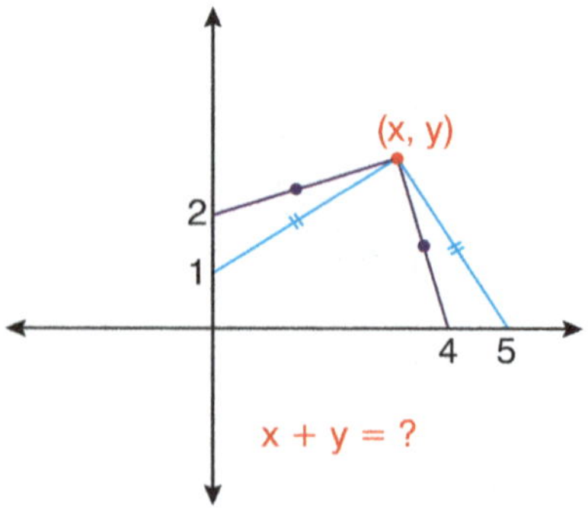

MathTopia Press

CHAPTER 6

TRIGONOMETRY CONNECTION

Definition 1.1

Consider a right triangle ABC with the right angle at C, let $\alpha = \angle A$.

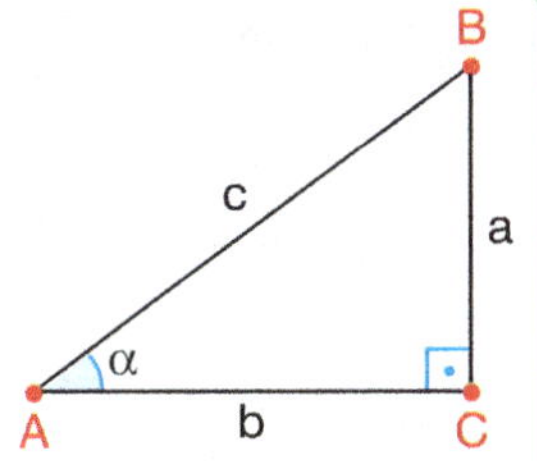

Label the sides as follows:

- Opposite side to α: $a = BC$
- Adjacent side to α: $b = AC$
- Hypotenuse: $c = AB$

Then the six trigonometric ratios of angle α are defined as:

$$\sin\alpha = \frac{a}{c}, \quad \cos\alpha = \frac{b}{c}, \quad \tan\alpha = \frac{a}{c},$$

and their reciprocals:

$$\csc\alpha = \frac{c}{a}, \quad \sec\alpha = \frac{c}{b}, \quad \cot\alpha = \frac{b}{a}.$$

Property 1.1

The unit circle is a circle centered at the origin with radius 1. Its equation is:

$$x^2 + y^2 = 1.$$

In trigonometry, we let

$$x = \cos\theta \text{ and } y = \sin\theta.$$

Substituting into the circle equation gives:

$$\sin^2\theta + \cos^2\theta = 1.$$

This fundamental relationship leads to several important trigonometric identities. Dividing both sides by $\cos^2\theta$ or $\sin^2\theta$, we obtain:

$$\sin^2\theta + \cos^2\theta = 1,$$
$$\tan^2\theta + 1 = \sec^2\theta,$$
$$\cot^2\theta + 1 = \csc^2\theta.$$

These ratios express the fundamental relationships between the sides of a right triangle and the acute angle α. They also satisfy the Pythagorean identities:

$$\sin^2\alpha + \cos^2\alpha = 1, \quad \tan\alpha \cdot \cot\alpha = 1,$$
$$\sec\alpha \cdot \cos\alpha = 1, \quad \csc\alpha \cdot \sin\alpha = 1.$$

MathTopia Press

Example 1.1

In ΔABC, suppose

$$\cos(2A - B) + \sin(A + B) = 2$$

and $AB = 4$. Find BC.

Solution:

Maximum occurs only if

$\cos(2A - B) = 1$ and $\sin(A + B) = 1$.

Thus,

$2A - B = 0 \Rightarrow B = 2A$, $A + B = 90° \Rightarrow 3A = 90° \Rightarrow A = 30°$, $B = 60°$, $C = 90°$.

In a $30° - 60° - 90°$ triangle,

$$\frac{BC}{AB} = \frac{1}{2}.$$

Given $AB = 4$,

$$BC = 2.$$

Definition 1.2

Trigonometric Symmetry and Co-function Identities

The trigonometric functions exhibit specific symmetry and co-function relationships based on angle reflection and complementary properties:

$$\sin(-\theta) = -\sin\theta, \cos(-\theta) = \cos\theta.$$

$\sin\theta = \cos(90° - \theta)$,

$\sin\theta = \sin(180° - \theta)$,

$\cos\theta = -\cos(180° - \theta)$

Summary:

- **Sine** is an **odd** function: its changes sign when the angle is negated.
- **Cosine** is an **even** function: it remains the same when the angle is negated.
- **Sine and cosine** are co-functions of complementary angles.
- **Sine** is symmetric in the second quadrant, while **cosine** changes sign there.

Example 1.2

Let x be a real number such that

$$\sec x - \tan x = 2.$$

Find the value of $\sec x + \tan x$.

Solution:

We use the Pythagorean identity:

$$\tan^2 x + 1 = \sec^2 x.$$

Rearranging gives

$$1 = \sec^2 x - \tan^2 x.$$

This can be factored as

$$1 = (\sec x - \tan x)(\sec x + \tan x).$$

Substituting $\sec x - \tan x = 2$, we get

$$1 = 2(\sec x + \tan x).$$

Therefore,

$$\sec x + \tan x = \frac{1}{2}.$$

Theorem 1.1

For any real numbers x and y:

1. Sine of a Sum and Difference

$$\sin(x + y) = \sin x \cos y + \cos x \sin y,$$
$$\sin(x - y) = \sin x \cos y - \cos x \sin y.$$

2. Cosine of a Sum and Difference

$$\cos(x + y) = \cos x \cos y - \sin x \sin y,$$
$$\cos(x - y) = \cos x \cos y + \sin x \sin y.$$

3. Tangent of a Sum and Difference

$$\tan(x + y) = \frac{\tan x + \tan y}{1 - \tan x \tan y},$$

$$\tan(x - y) = \frac{\tan x - \tan y}{1 + \tan x \tan y}.$$

Theorem 1.2

1. $\sin 2x = 2\sin x \cos x$
2. $\cos 2x = \cos^2 x - \sin^2 x = 2\cos^2 x - 1 = 1 - 2\sin^2 x$
3. $\tan 2x = \dfrac{2\tan x}{1 - \tan^2 x}$

MathTopia Press

Proof 1.1

(1) Sine double angle.

Use the sum formula:

$$\sin(2x) = \sin(x + x) = \sin x \cos x + \cos x \sin x = 2\sin x \cos x.$$

(2) Cosine double angle.

Start from the sum formula:

$$\cos(2x) = \cos(x + x) = \cos^2 x - \sin^2 x.$$

Now apply $\sin^2 x + \cos^2 x = 1$:

$$\cos^2 x - \sin^2 x = (1 - \sin^2 x) - \sin^2 x = 1 - 2\sin^2 x,$$

or

$$\cos^2 x - \sin^2 x = \cos^2 x - (1 - \cos^2 x) = 2\cos^2 x - 1.$$

(3) Tangent double angle.

Method A (from sine / cosine):

$$\tan(2x) = \frac{\sin(2x)}{\cos(2x)} = \frac{2\sin x \cos x}{\cos^2 x - \sin^2 x} = \frac{2(\sin x / \cos x)}{1 - (\sin x / \cos x)^2}$$

$$= \frac{2\tan x}{1 - \tan^2 x}.$$

Method B (direct sum formula for tan):

$$\tan(2x) = \tan(x + x) = \frac{\tan x + \tan x}{1 - \tan x \tan x} = \frac{2\tan x}{1 - \tan^2 x}.$$

Example 1.3

Find $\sin 75°$.

Solution:

Using the sum formula for sine:

$$\sin(75°) = \sin(30° + 45°)$$
$$= \sin 30° \cos 45° + \cos 30° \sin 45°$$
$$= \left(\frac{1}{2}\right)\left(\frac{\sqrt{2}}{2}\right) + \left(\frac{\sqrt{3}}{2}\right)\left(\frac{\sqrt{2}}{2}\right)$$
$$= \frac{\sqrt{6} + \sqrt{2}}{4}.$$

Example 1.4

Given

$$\sin + \sin b = \sqrt{\frac{5}{3}},\ \cos a + \cos b = 1,$$

find $\cos(a - b)$.

Solution:

$$(\sin a + \sin b)^2 + (\cos a + \cos b)^2 = \frac{5}{3} + 1 = \frac{8}{3}.$$

Expand LHS:

$$\sin^2 a + \sin^2 b + 2\sin a \sin b + \cos^2 a + \cos^2 b + 2\cos a \cos b$$

$$= 2 + 2(\sin a \sin b + \cos a \cos b).$$

But $\sin a \sin b + \cos a \cos b = \cos(a - b)$.

So $2 + 2\cos(a - b) = \frac{8}{3} \Rightarrow \cos(a - b) = \frac{1}{3}$.

$$\cos(a - b) = \frac{1}{3}.$$

Example 1.5

If

$$\sin x + \cos x = \frac{1}{2},$$

find the value of

$$\sin^3 x + \cos^3 x.$$

Solution:

Use the identity:

$$\sin^3 x + \cos^3 x = (\sin x + \cos x)(\sin^2 x - \sin x \cos x + \cos^2 x).$$

Since $\sin 2x + \cos 2x = 1$,

$$\sin^3 x + \cos^3 x = (\sin x + \cos x)(1 - \sin x \cos x).$$

Now square the given equation:

$$(\sin x + \cos x)^2 = \frac{1}{4} \Rightarrow 1 + 2\sin x \cos x = \frac{1}{4}.$$

So,

$$\sin x \cos x = -\frac{3}{8}$$

Then,

$$\sin 3x + \cos 3x = \frac{1}{2}(1 - (-\frac{3}{8})) = \frac{1}{2} \cdot \frac{11}{8} = \frac{11}{16}.$$

Theorem 1.3

For any real numbers x and y, following identities hold:

1. $\sin x \sin y = \frac{1}{2}[\cos(x - y) - \cos(x + y)]$,

2. $\cos x \cos y = \frac{1}{2}[\cos(x + y) + \cos(x - y)]$,

3. $\sin x \cos y = \frac{1}{2}[\sin(x + y) + \sin(x - y)]$.

Proof 1.2

(1)

Start with the cosine addition and subtraction formulas:

$$\cos(x - y) = \cos x \cos y + \sin x \sin y,$$

$$\cos(x + y) = \cos x \cos y - \sin x \sin y.$$

Subtract the second from the first:

$$\cos(x - y) - \cos(x + y) = 2\sin x \sin y.$$

Dividing both sides by 2 gives:

$$\sin x \sin y = \frac{1}{2}[\cos(x - y) - \cos(x + y)].$$

(2)

Adding the same two cosine formulas gives:

$$\cos(x - y) + \cos(x + y) = 2\cos x \cos y.$$

Hence,

$$\cos x \cos y = \frac{1}{2}[\cos(x + y) + \cos(x - y)].$$

(3)

From the sine addition and subtraction formulas:

$$\sin(x + y) = \sin x \cos y + \cos x \sin y,$$

$$\sin(x - y) = \sin x \cos y - \cos x \sin y.$$

Adding these gives:

$$\sin(x + y) + \sin(x - y) = 2\sin x \cos y.$$

Thus,

$$\sin x \cos y = \frac{1}{2}[\sin(x + y) + \sin(x - y)].$$

Theorem 1.4

For any triangle ΔABC with circumradius R:

$$\frac{a}{\sin A} = \frac{b}{\sin B} = \frac{c}{\sin C} = 2R.$$

Proof 1.3

Let the circumcircle of ΔABC have center O and radius R.

Draw the perpendicular from O to side BC, meeting it at X.

Because $OB = OC = R$ and $OX \perp BC$, triangle BOX is right-angled at X.

Hence,

$$\sin A = \frac{BX}{R}.$$

But since $BX = \frac{a}{2}$ (as X is the midpoint of chord BC),

$$\sin A = \frac{a}{2R}.$$

Therefore,

$$\frac{a}{\sin A} = 2R.$$

By symmetry,

$$\frac{b}{\sin B} = \frac{c}{\sin C} = 2R.$$

Theorem 1.4

For any triangle ΔABC:

$$c^2 = a^2 + b^2 - 2ab\cos(C)$$

Proof 1.4

Drop a perpendicular from vertex B to side AC, meeting it at X.

Then $AX = c\cos(B)$ and $CX = a - c\cos(B)$.

From right triangle BXC:

$$b^2 = (a - c\cos B)^2 + (c\sin B)^2.$$

Expanding and simplifying gives:

$$b^2 = a^2 + c^2 - 2ac\cos B.$$

By symmetry, we can write:

$$c^2 = a^2 + b^2 - 2ab\cos C.$$

MathTopia Press

Theorem 1.5

Prove that the area of a quadrilateral is equal to half the product of its diagonals and the sine of the included angle.

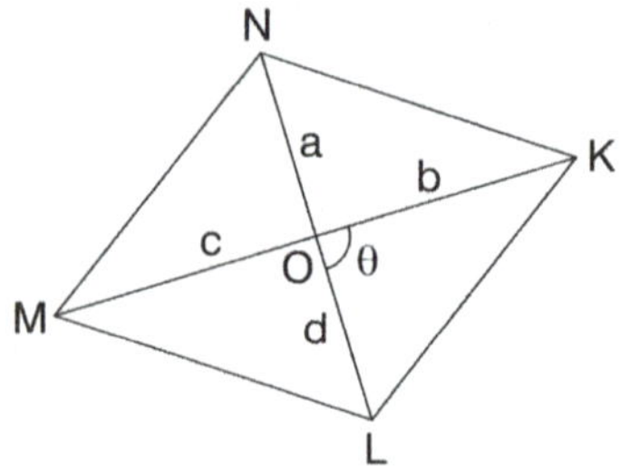

Proof 1.5

Let diagonals NL and MK intersect at O with angle θ.

Set $ON = a$, $OK = b$, $OL = d$, $OM = c$.

Then

$$|MNKL| = \frac{1}{2}\sin\theta\,(ab + bd + dc + ca)$$

$$= \frac{1}{2}\sin\theta\,(a + d)\,(b + c).$$

Since $NL = a + d$ and $MK = b + c$,

$$|MNKL| = \frac{1}{2}NL \cdot MK \cdot \sin\theta.$$

Theorem 1.6

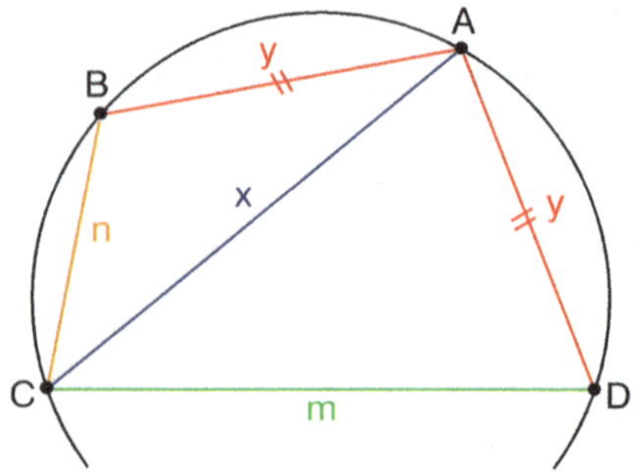

ABCD is a cyclic quadrilateral such that $AB = AD$.

Prove that: $(x + y)\,(x - y) = mn$

Proof 1.6

Let $\angle CAB = \theta$.

By the Cosine Law in triangles ABC and ADC:

$$x^2 = n^2 + y^2 - 2ny\cos\theta \ (1)$$

and

$$x^2 = m^2 + y^2 - 2my\cos(\pi - \theta). \ (2)$$

Since $\cos(\pi - \theta) = -\cos\theta$,

$$x^2 = m^2 + y^2 + 2my\cos\theta. \ (2)$$

Multiply (1) by m and (2) by n:

$$mx^2 = mn^2 + my^2 - 2mny\cos\theta, \ (A)$$

$$nx^2 = nm^2 + ny^2 + 2mny\cos\theta. \ (B)$$

Add (A) and (B):

$$(m + n)x^2 = mn(m + n) + (m + n)y^2.$$

Simplify by dividing through by $(m + n) \neq 0$:

$$x^2 = mn + y^2.$$

Thus,

$$x^2 - y^2 = mn,$$

which gives

$$(x + y)(x - y) = mn.$$

Example 1.6

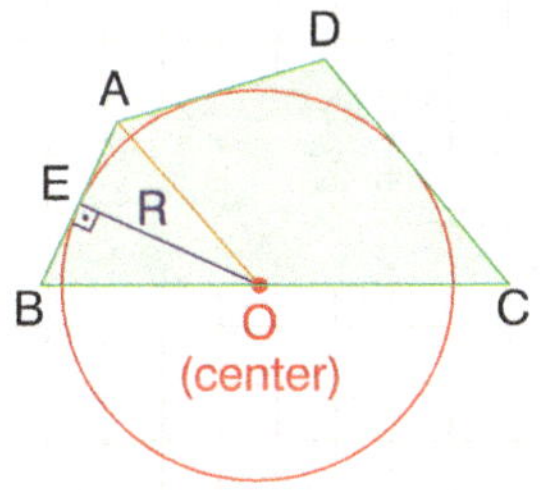

In cyclic quadrilateral ABCD,

$\angle BAD + \angle BCD = \pi$.

Find $\dfrac{AB + CD}{BC}$.

Solution:

Let the circle have radius R.

From the extended sine rule in a triangle:

side = 2Rsin (opposite angle).

So, for triangles sharing sides of the quadrilateral:

$$AB = 2R\sin C, \quad CD = 2R\sin B.$$

The chord BC subtends angles B and C at the circum ference, so

$$BC = 2R(\sin B + \sin C).$$

Calculation

$$\frac{AB + CD}{BC} = \frac{2R(\sin B + \sin C)}{2R(\sin B + \sin C)} = 1.$$

Example 1.7

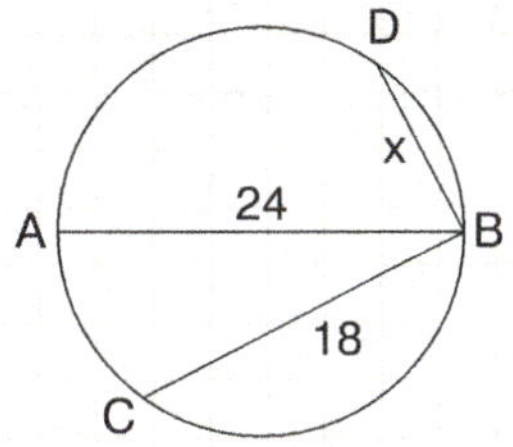

In the semicircle with diameter AB,

$m(\angle ABD) = 2 \times m(\angle ABC)$

and

AB = 24 cm, BC = 18 cm.

Find the length BD = x.

Solution:

- AB is a diameter $\Rightarrow \angle ABC$ is right at C.

  So $AB^2 = AC^2 + BC^2 \Rightarrow 24^2 = AC^2 + 18^2 \Rightarrow AC = 6\sqrt{7}$.
- Let $\alpha = \angle ABC$. Since α subtends arc AC, with radius R = 12.

  $AC = 2R\sin(\angle AOC / 2) = 24\sin\alpha = 6\sqrt{7} \Rightarrow \sin\alpha = \dfrac{\sqrt{7}}{4}$.
- Given $\angle ABD = 2\alpha$. The chord BD corresponds to arc BD whose central angle is

  $\angle BOD = 180° - 4\alpha \Rightarrow BD = 2R\sin(\dfrac{\angle BOD}{2})$

$= 24\sin(90° - 2\alpha)$.

- With $\sin\alpha = \dfrac{\sqrt{7}}{4}$,

  $\cos(2\alpha) = 1 - 2\sin^2\alpha = 1 - 2 \cdot \dfrac{7}{16} = \dfrac{1}{8}$.

Thus $BD = 24 \cdot \dfrac{1}{8} = 3$.

MathTopia Press

My Notes:

Example 1.8

In square ABCD (side = 28), point M lies on BC with BM = 21, and point N lies CD with DN = 4 (so NC = 24). Find $\angle MAN$.

Solution:

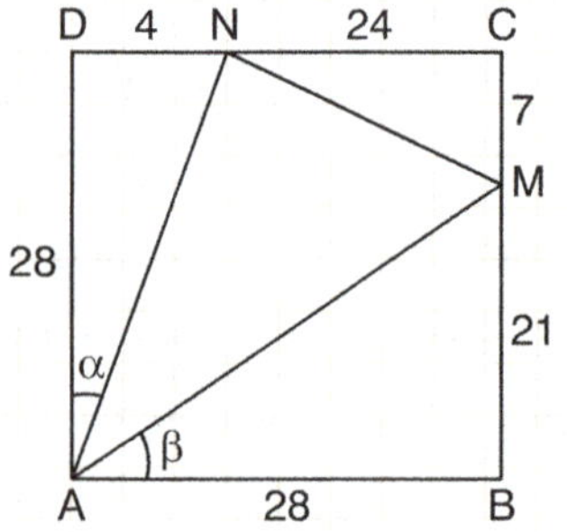

Let $\alpha = \angle DAN$ and $\beta = \angle MAB$. Then $\angle MAN = \alpha + \beta$.

Because N is 4 units from D along the top side, the right triangle with vertical 28 and horizontal 4 gives

$$\tan\alpha = \frac{4}{28}.$$

Similarly, since M is 21 units above B,

$$\tan\beta = \frac{21}{28}.$$

Thus

$$\tan(\alpha + \beta) = \frac{\tan\alpha + \tan\beta}{1 - \tan\alpha \cdot \tan\beta} = \frac{\frac{4}{28} + \frac{21}{28}}{1 + \frac{4}{28} \cdot \frac{21}{28}}$$

$$= \frac{25 / 28}{1 - 84 / 784} = \frac{25 / 28}{700 / 784} = 1.$$

Hence $\alpha + \beta = 45°$. Therefore

$$\angle MAN = 45°.$$

My Notes:

Example 1.9

In triangle ABC, let D be the foot of the altitude from C to AB, and let H be the orthocenter. Given CH = HD, find $\tan A \cdot \tan B$.

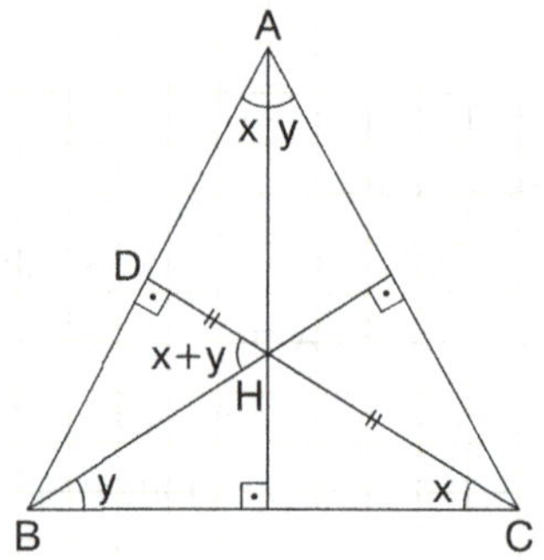

Solution:

Let $\angle C = x$, $\angle B = y$, so $\angle A = x + y$.

1. From the right-triangle geometry at the feet of the altitudes (see figure), the angle between AH and HD equals $x + y$.

$$\tan A = \tan(x + y) = \frac{DB}{HD}.$$

2. Because CH = HD, the right triangles based at D and at the foot of the altitude from B are congruent about H; hence the projection on AB doubles, giving.

$$\tan B = \frac{2HD}{DB}.$$

3. Multiply:

$$\tan A \cdot \tan B = \frac{DB}{HD} \cdot \frac{2HD}{DB} = 2.$$

So $\tan A \tan B = 2$.

My Notes:

1.

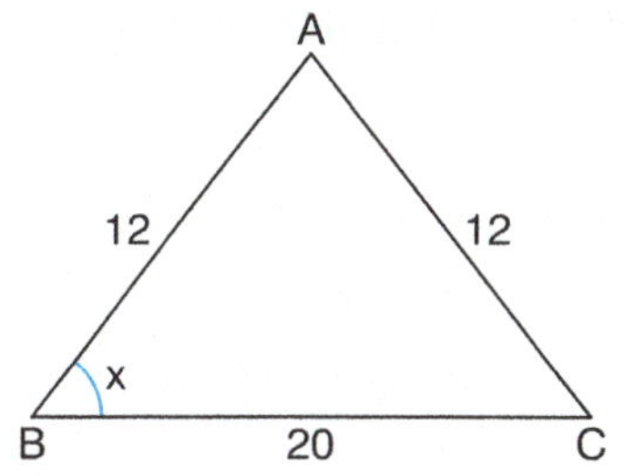

In isosceles triangle ABC

m(∠ABC) = x

|AB| = |AC| = 12 cm

|BC| = 20 cm

Given the above, what is secx?

2.

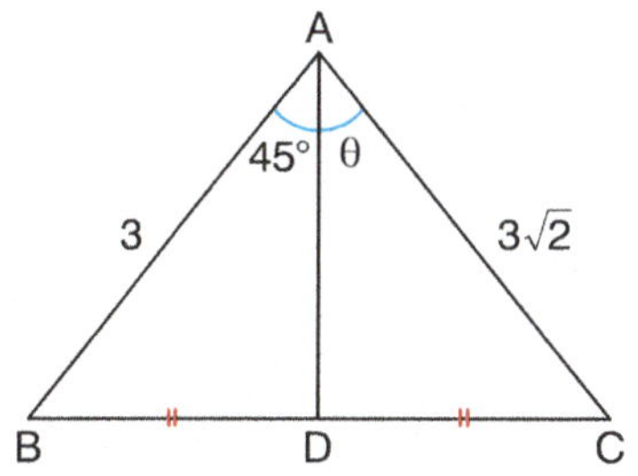

In triangle ABC

|BD| = |DC|

m(∠BAD) = 45°

m(∠DAC) = θ

|AB| = 3 cm

|AC| = $3\sqrt{2}$ cm

Given the above, what is sinθ?

3. If triangle ABC has AB = 4, AC = 8, and ∠C = 30°, find the area of the triangle ABC.

4. If $\sin(2x + y) = 2\sin y$, find $\dfrac{\tan(x + y)}{\tan x}$.

5. Let ABCD be a square.

- The side length of the square is |AB| = 6.
- Point E lies on side DC such that |DE| = 2.
- Point F lies on side BC such that |BF| = 3.

Draw the square and place points E and F accordingly. Find the measure of ∠EAF.

6. In a triangle ABC, the following conditions are given:

- ∠B = 2∠A = 2α.
- Side AC = 12.
- Side BC = 8.

Find cosα.

7.

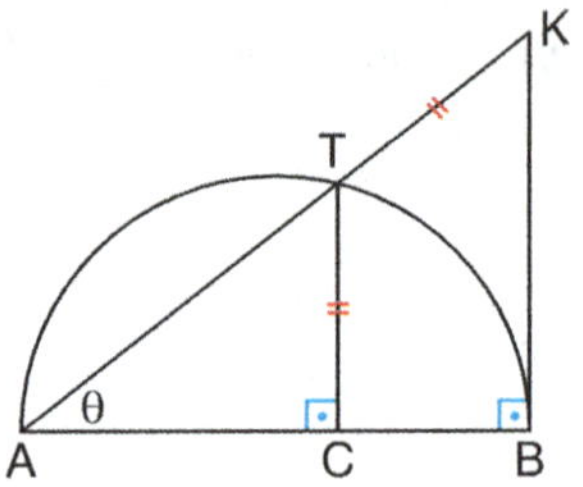

$|AB| = 1$

$|TK| = |TC| \rightarrow \sin\theta = ?$

8. (2012 AMC 12A)

A triangle has area 30, one side of length 10, and the median to that side of length 9. Let θ be the acute angle formed by that side and median. What is $\sin\theta$?

9.

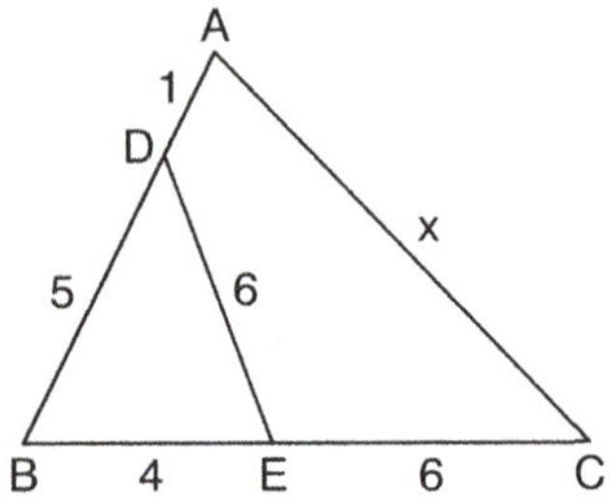

In triangle ABC:

- AD = 1, DB = 5, so AB = 6.
- BE = 4, EC = 6, so BC = 10.
- DE = 6.

Find the length of AC = x.

10. What is the Area of region shaded in blue?

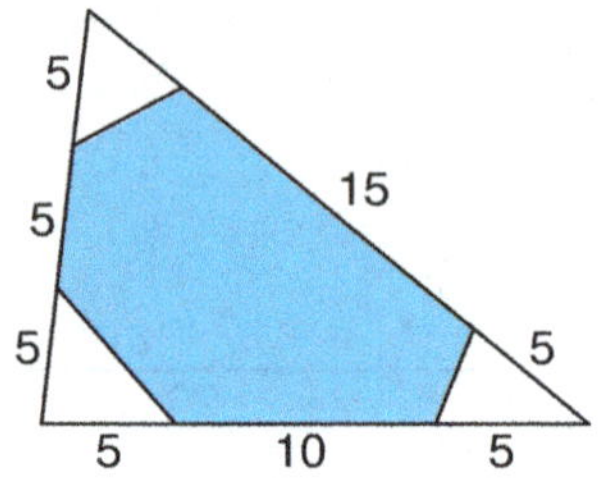

11. (2014 ARML 4)

In triangle ABC, a = 12, b = 17, and c = 13. Compute $b\cos C - c\cos B$.

12.

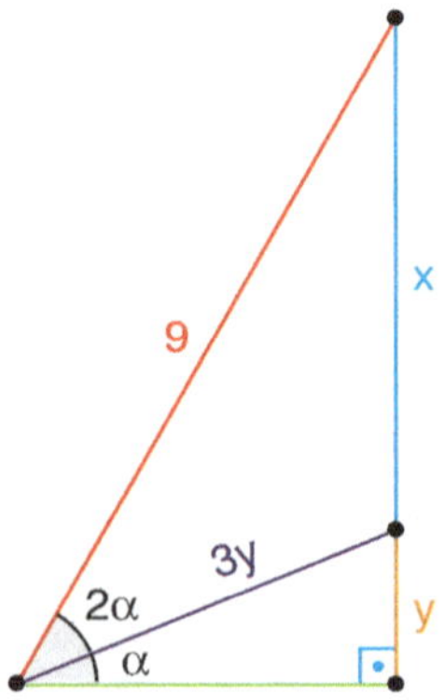

Find the value of x?

MathTopia Press

1. In triangle $\triangle ABC$, point D is the foot of the A-angle bisector. If $AD = 6$, $BD = 4$, and $CD = 3$, find AB.

2. ABCD is a square, $E \in \overline{BC}$, $F \in \overline{CD}$, $BE = 2CE$, and $m\angle FAB = m\angle EFA$. What is the value of $\tan\angle DAF$?

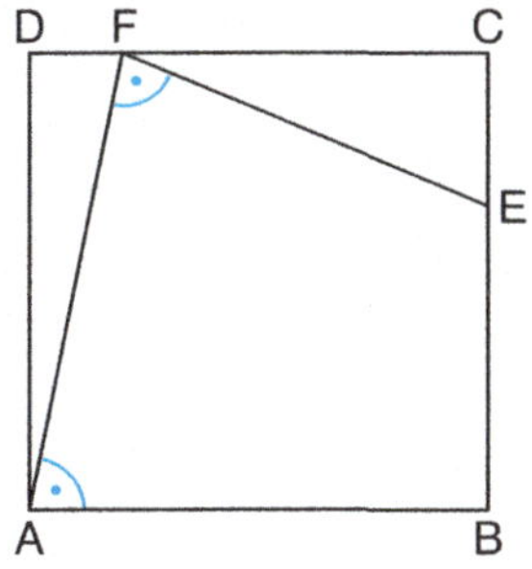

3. $(3 - k)x + (2k - 1)y - 3 = 0$

The angle that the line above makes with the x-axis in the positive direction is α.

If $\tan\alpha = \frac{2}{3}$, what is k?

4. Area(ABC) = ?

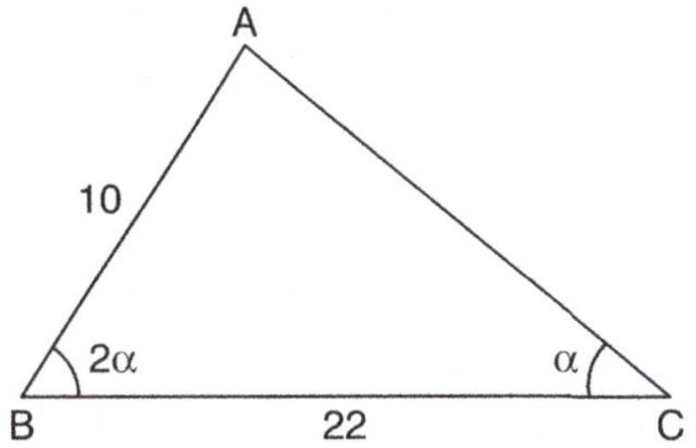

5. Area(ABC) = ?

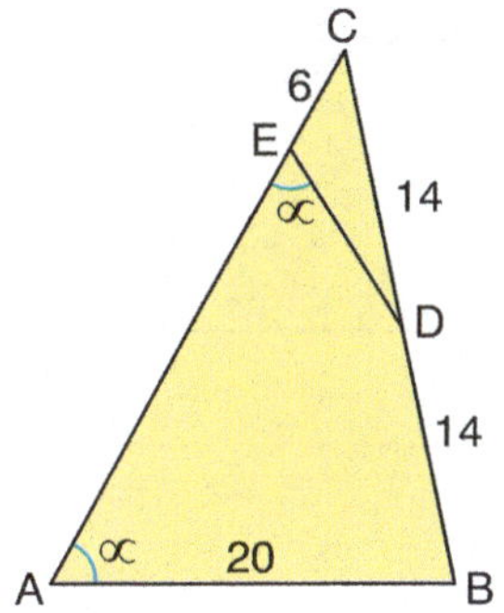

6. In a scalene triangle ABC, the side lengths that are opposite of the angles A, B, C are a units, b units and c units respectively.

$$a^2 = \frac{a^3 + b^3 - c^3}{a + b - c}$$

Given the above, what is the measure of the angle A in degrees?

7. Suppose the angles x and y satisfy

$$\sin(2x + y) = 7\sin y.$$

Find the value of

$$\frac{\tan(x + y)}{\tan x}.$$

8. Point O is the center of the semicircle; point C is the center of the quarter circle. Given that T is the tangent point and m(∠ABT) = x, find the value of sin(x).

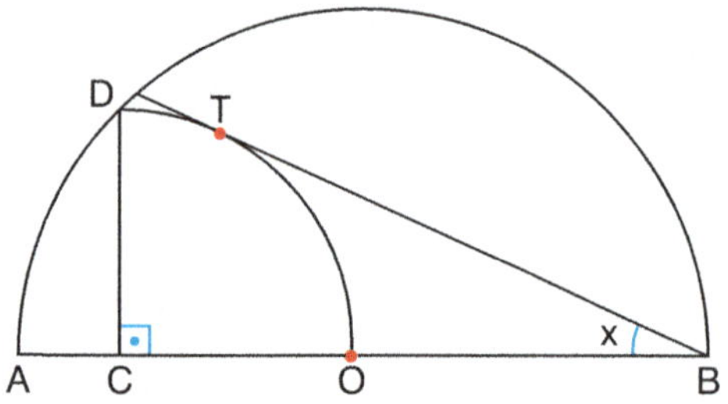

9.

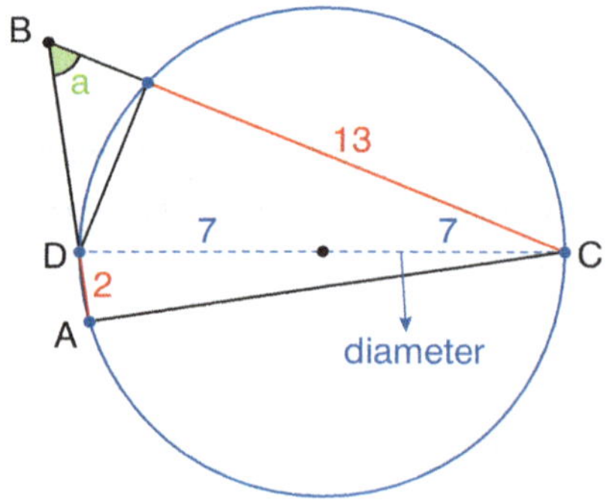

m∠ABC = ?

10. ABCD is square.

- |DE| = 3 units
- |FC| = 1 unit
- |CE| = 2 units

Find tan(x).

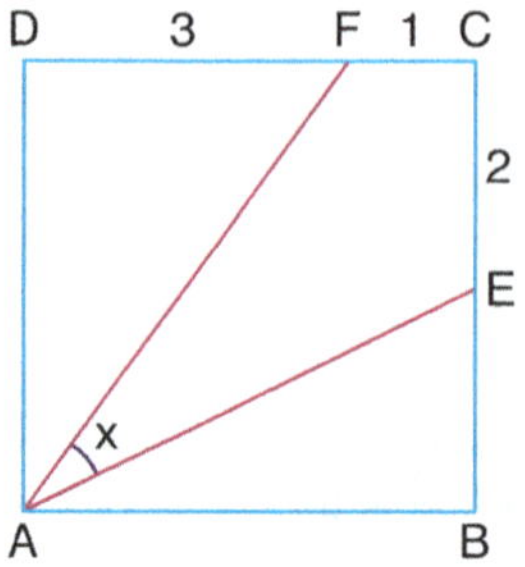

11.

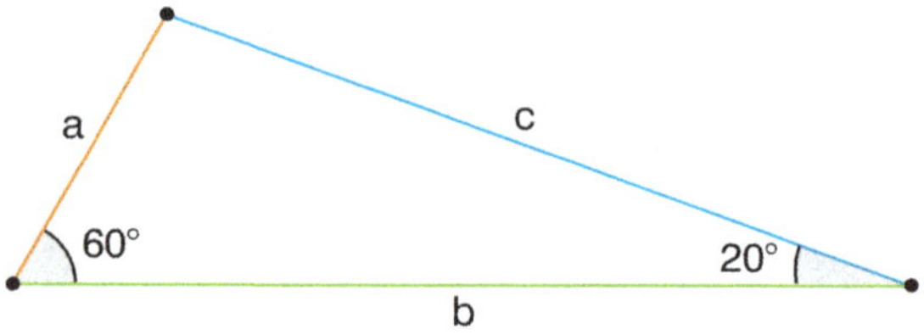

$$\frac{b}{a} - \frac{c}{b} = ?$$

12. In trapezoid ABCD with AB || CD, a point E is chosen on side BC. Suppose BE > EC. The areas of the triangles satisfy

|ABE| = 15, |AED| = 23, |ECD| = 4.

What is the ratio $\frac{BE}{EC}$?

1. **(2019 AMC 12A, #12)**

In ΔABC with integer side lengths,

$\cos A = \frac{11}{16}$, $\cos B = \frac{7}{8}$, and $\cos C = -\frac{1}{4}$.

What is the least possible perimeter for ΔABC?

A) 9 B) 12 C) 23 D) 27 E) 44

2. **(2019 AMC 12B, #12)**

Right triangle ACD with right angle at C is constructed outwards on the hypotenuse $\overline{AC}$ of isosceles right triangle ABC with leg length 1, as shown, so that the two triangles have equal perimeters.

What is $\sin(2\angle BAD)$?

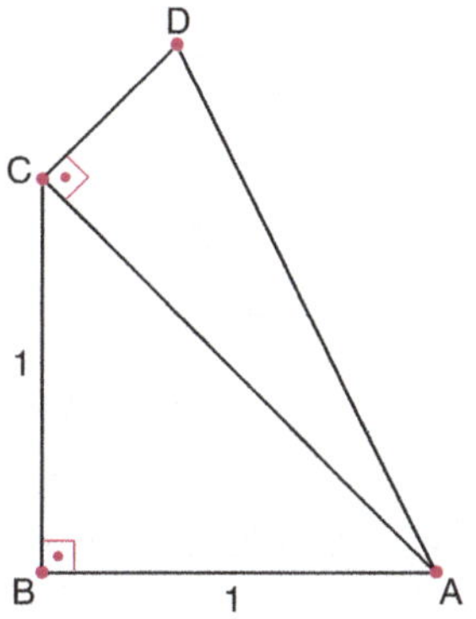

A) $\frac{1}{3}$ B) $\frac{\sqrt{2}}{2}$ C) $\frac{3}{4}$ D) $\frac{7}{9}$ E) $\frac{\sqrt{3}}{2}$

3. **(2000 AMC 12, #17)**

A circle centered at O has radius 1 and contains the point A. The segment AB is tangent to the circle at A and $\angle AOB = \theta$.

If point C lies on $\overline{OA}$ and $\overline{BC}$ bisects $\angle ABO$, then OC = ?

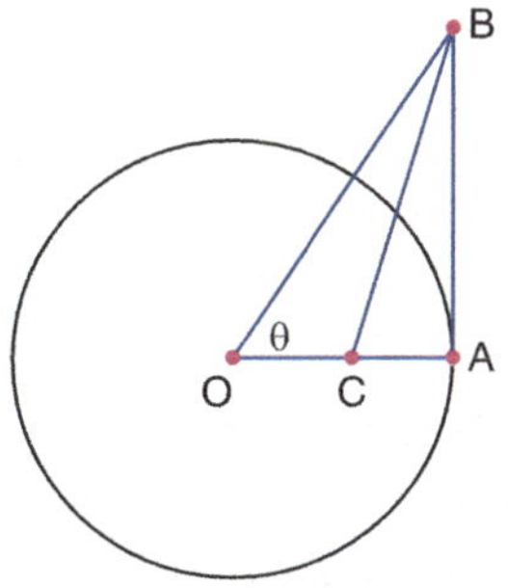

A) $\sec^2\theta - \tan\theta$ B) $\frac{1}{2}$ C) $\frac{\cos^2\theta}{1 + \sin\theta}$

D) $\frac{1}{1 + \sin\theta}$ E) $\frac{\sin\theta}{\cos^2\theta}$

4. **(MathCON 2019 Final Gr10, #24)**

In the circle with center O and diameter $\overline{AB}$, suppose $\overline{BC} \perp \overline{AB}$, $\overline{DE} \perp \overline{AB}$, and $m\angle OCB = x$.

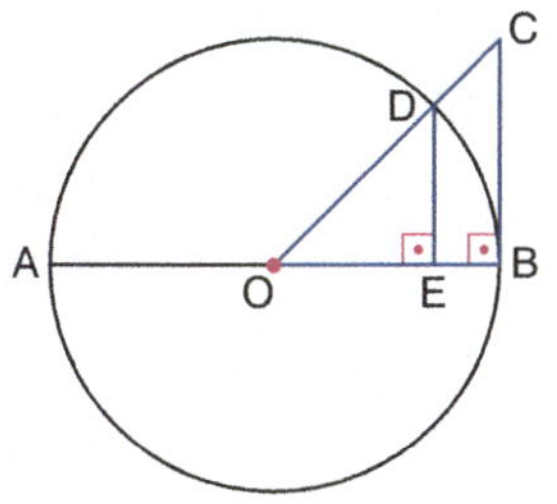

What is the trigonometric value of

$\frac{BC + DC}{DE + BE}$?

A) sin x B) tan x C) cot x D) csc x E) sec x

5. **(2007 AMC 12A, #17)**

Suppose that $\sin a + \sin b = \sqrt{\frac{5}{3}}$ and $\cos a + \cos b = 1$.

What $\cos(a - b)$?

A) $\sqrt{\frac{5}{3}} - 1$ B) $\frac{1}{3}$ C) $\frac{1}{2}$ D) $\frac{2}{3}$ E) 1

6. **(2007 AMC 12B, #19)**

Rhombus ABCD, with side length 6, is rolled to form a cylinder of volume 6 by taping $\overline{AB}$ to $\overline{DC}$.

What is $\sin(\angle ABC)$?

A) $\frac{\pi}{9}$ B) $\frac{1}{2}$ C) $\frac{\pi}{6}$ D) $\frac{\pi}{4}$ E) $\frac{\sqrt{3}}{2}$

7. (2011 AMC 12B, # 14)

A segment through the focus F of a parabola with vertex V is perpendicular to $\overline{FV}$ and intersects the parabola in A and B.

What is cos(∠AVB)?

A) $-\frac{3\sqrt{5}}{7}$ B) $-\frac{3\sqrt{5}}{7}$ C) $-\frac{4}{5}$ D) $-\frac{3}{5}$ E) $-\frac{1}{2}$

8. (2009 AMC 12A, #13)

A ship sails 10 miles in a straight line from A to B, turns through an angle between 45° and 60°, and then sails another 20 miles to C. Let AC be measured in miles.

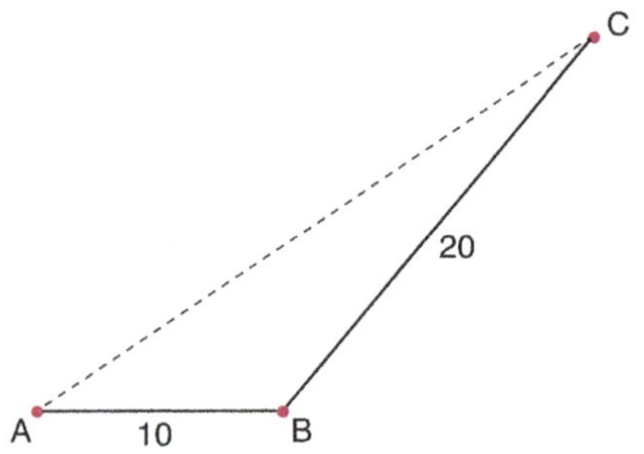

Which of the following intervals contains AC^2?

A) [400, 500] B) [500, 600] C) [600, 700] D) [700, 800] E) [800, 900]

9. (2010 AMC 12B, #13)

In ΔABC, cos(2A – B) + sin(A + B) = 2 and AB = 4.

What is BC?

A) $\sqrt{2}$ B) $\sqrt{3}$ C) 2 D) $2\sqrt{2}$ E) $2\sqrt{3}$

10. (2013 AMC 12B, #20)

For 135° < x < 180°, points

P = (cos x, $\cos^2$ x), Q = (cot x, $\cot^2$ x), R = (sin x, $\sin^2$ x)

and S = (tan x, $\tan^2$ x) are the vertices of a trapezoid.

What is sin(2x)?

A) $2-2\sqrt{2}$ B) $3\sqrt{3}-6$ C) $3\sqrt{2}-5$ D) $-\frac{3}{4}$ E) $1-\sqrt{3}$

11. (2010 AMC 12B, # 22)

Let ABCD be a cyclic quadrilateral.

The side lengths of ABCD are distinct integers less than 15 such that BC · CD = AB · DA.

What is the largest possible balue of BD?

A) $\sqrt{\frac{325}{2}}$ B) $\sqrt{185}$ C) $\sqrt{\frac{389}{2}}$ D) $\sqrt{\frac{425}{2}}$ E) $\sqrt{\frac{533}{2}}$

12. (2008 AMC 12A, #24)

Triangle ABC has ∠C = 60° and BC = 4.

Point D is the midpoint of BC.

What is the largest possible value of tan∠BAD?

A) $\frac{\sqrt{3}}{6}$ B) $\frac{\sqrt{3}}{2}$ C) $\frac{\sqrt{3}}{2\sqrt{2}}$ D) $\frac{\sqrt{3}}{4\sqrt{2}-3}$ E) 1

1.

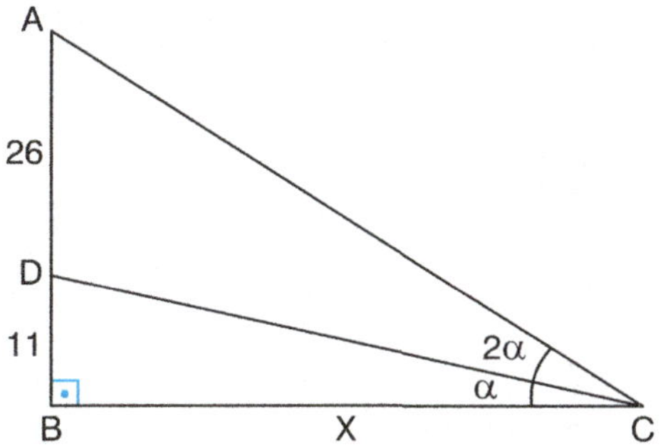

2. AB = CD = 6 cm

AD = ?

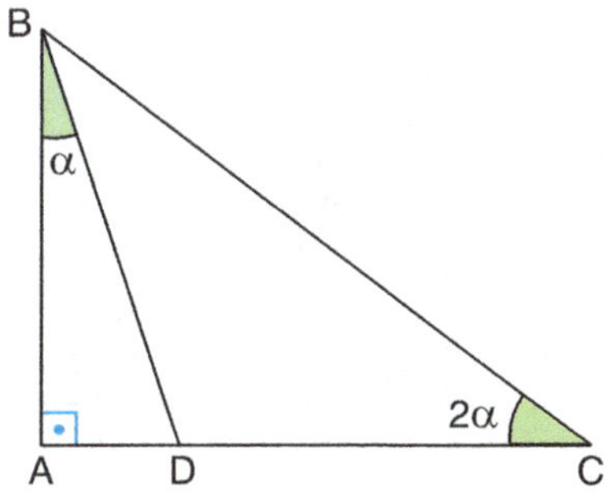

3. Green area = ?

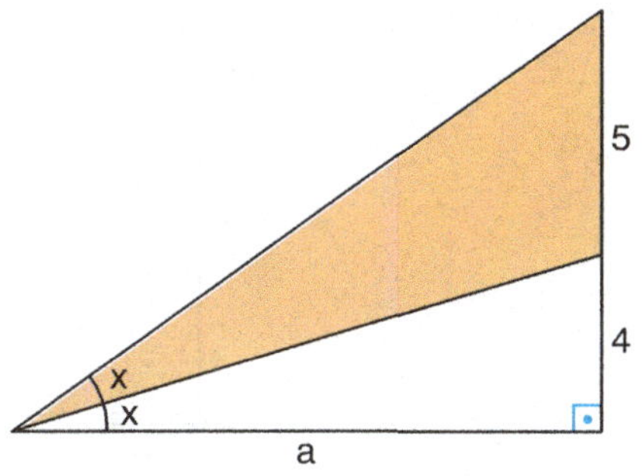

4. ABCD is a square.

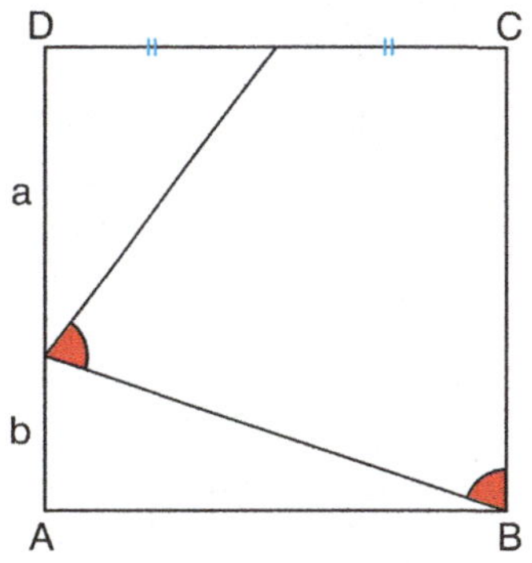

Find a/b

5.

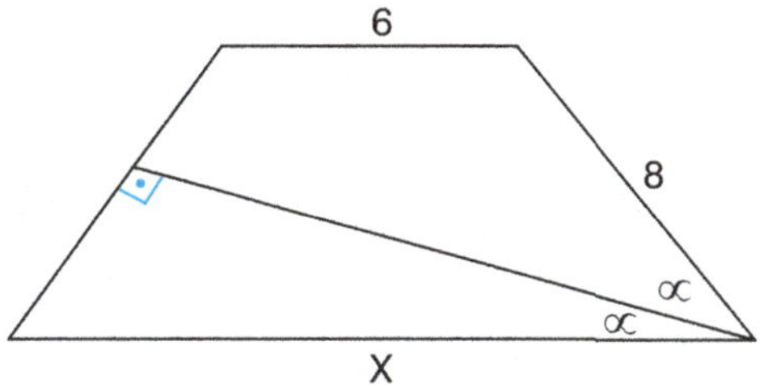

6.

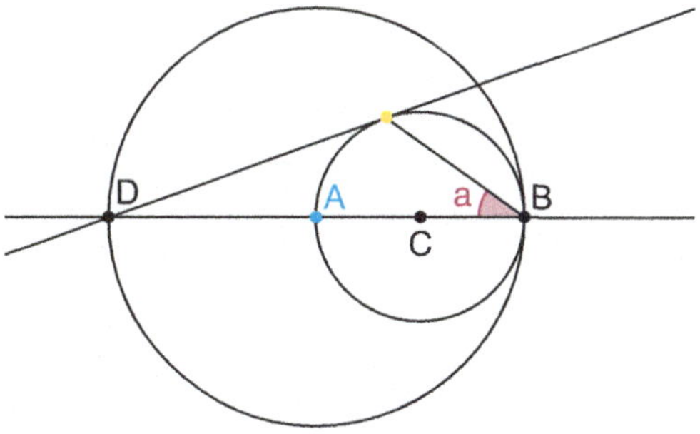

AB = AD

AC = BC

tan(a) ?

7. $\cot\alpha = ?$

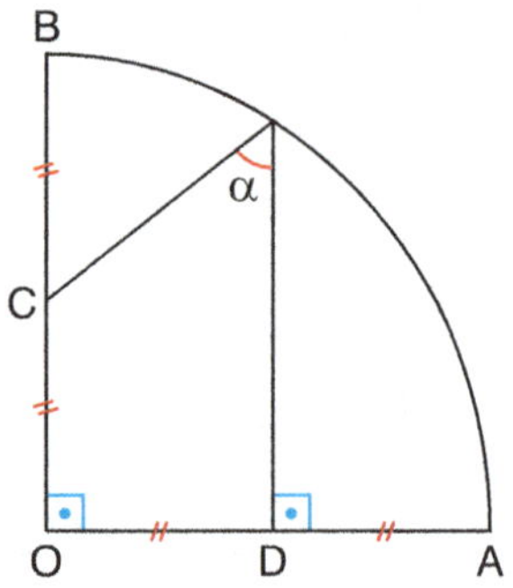

8.

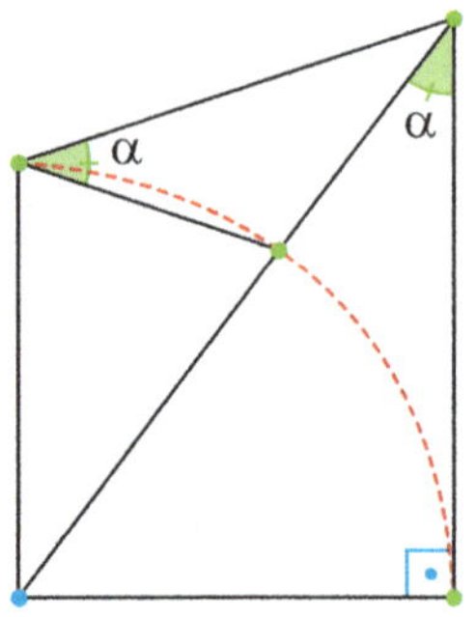

$\tan(\alpha) = ?$

9.

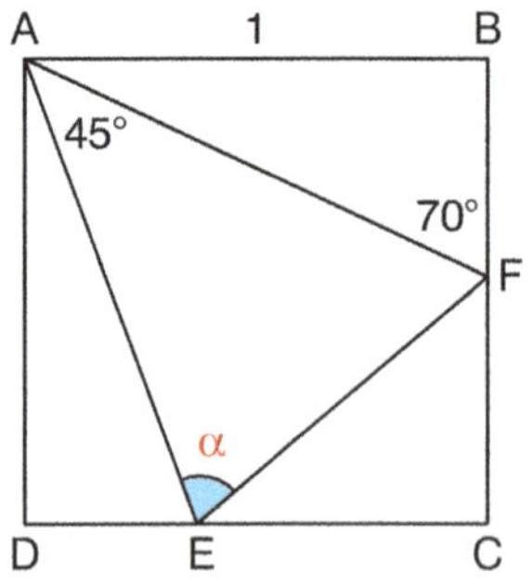

$\alpha = ?$

10.

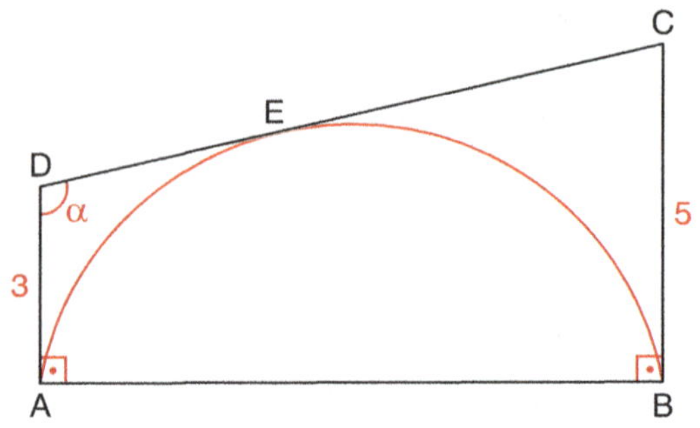

$\cos\alpha = ?$

11.

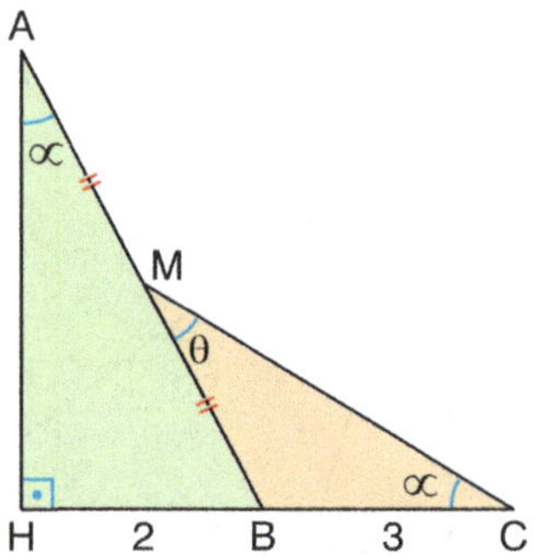

$\cot\alpha \cdot \tan\theta$

12.

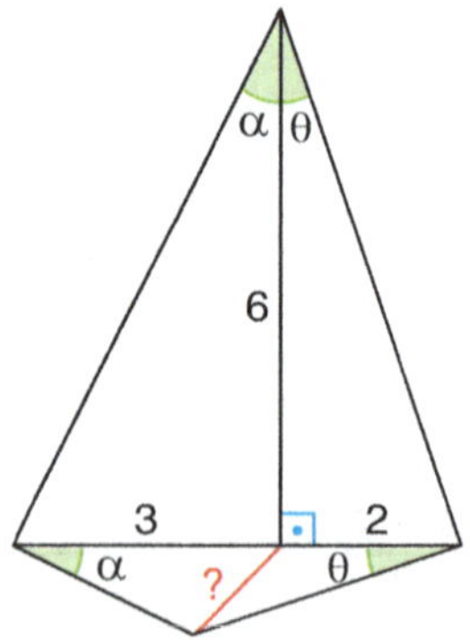

CHAPTER 7

3D Geometry

Definition 1.1

Base (upper/lower): The two congruent, parallel faces of a prism. They determine the prism's name (e.g., "triangular prism").

Basal edge: Any side (edge) of a base.

Lateral face: Any face that is **not** a base. Lateral faces connect the two bases.

Altitude (or height): A line segment that joins the two bases **and is perpendicular** to both. It represents the **distance between the bases**.

Face diagonal: A line segment joining **two vertices** that lie on the **same face**.

Space diagonal: A line segment joining **two vertices** that are **not on the same face**.

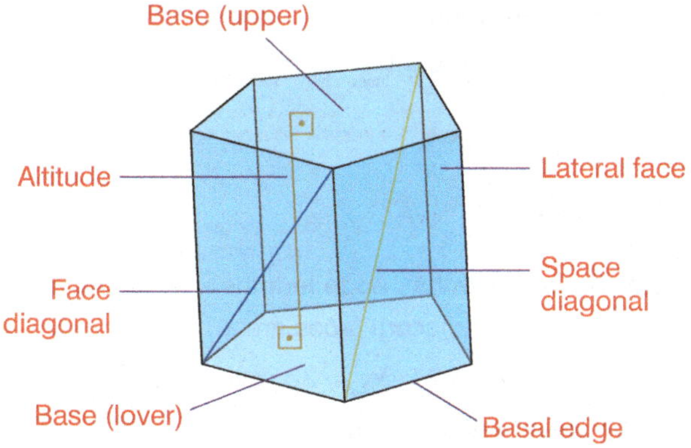

Definition 1.2

A **regular polyhedron** is a polyhedron whose faces are all congruent regular polygons.

There are five regular polyhedra, which are described below.

Name	Face Type	#Faces	#Edges	#Vertices
Tetrahedron	Triangle	4	6	4
Cube	Square	6	12	8
Octahedron	Triangle	8	12	6
Dodecahedron	Pentagon	12	30	20
Icosahedron	Triangle	20	30	12

Theorem 1.1

Euler's Polyhedral Formula

For any **convex polyhedron** with

- V = number of vertices,
- E = number of edges, and
- F = number of faces,

the relationship is:

$$V + F = E + 2.$$

Property 1.1

An n-gonal prism has $n + 2$ faces, $2n$ vertices, and $3n$ edges.

It contains $n(n - 1)$ face diagonals and $n(n - 3)$ space diagonals.

For example, a hexagonal prism ($n = 6$) has 8 faces, 12 vertices, 18 edges, 30 face diagonals, and 18 space diagonals.

Example 1.1

A prism has 21 edges. Find the number of faces and vertices of this prism.

Solution:

If a prism has 21 edges, this number is three times the number of vertices at its base. So the prism is a heptagonal prism, i.e, its base has 7 sides. So it has $7 + 2 = 9$ faces and $7 \cdot 2 = 14$ vertices.

Example 1.2

A prism has 18 vertices. Find how many space diagonals it has.

Solution:

For an n-gonal prism, the number of vertices is 2n.

So $2n = 18 \Rightarrow n = 9$.

Using the formula for space diagonals, $n(n -3)$:

$$9(9 - 3) = 54$$

Definition 1.3

A cuboid (also called a rectangular solid) is a special type of parallelepiped. It is parallelepiped whose faces are all rectangles. In a cuboid, all the faces join each other at right angles. Four edges which meet at a common vertex of a cuboid form the dimensions (a, b and c) of the cuboid.

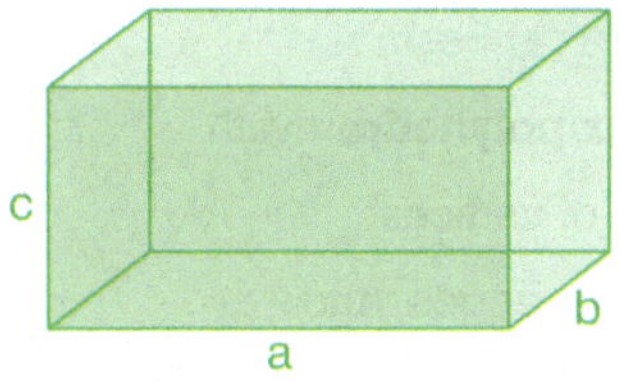

Definition 1.4

A cube (or regular hexahedron) is a prism whose six faces are congruent squares. It can also be described as a right square prism with equal edges, equal faces, and equal space diagonals.

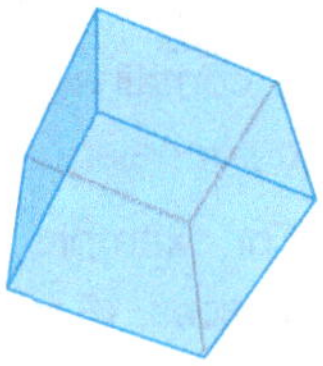

There are 11 possible nets for a cube–four are shown here. Can you find the other seven?

Example 1.3

Find the area of triangle ACD' in the adjacent cube if the edge length is $6\sqrt{2}$ cm.

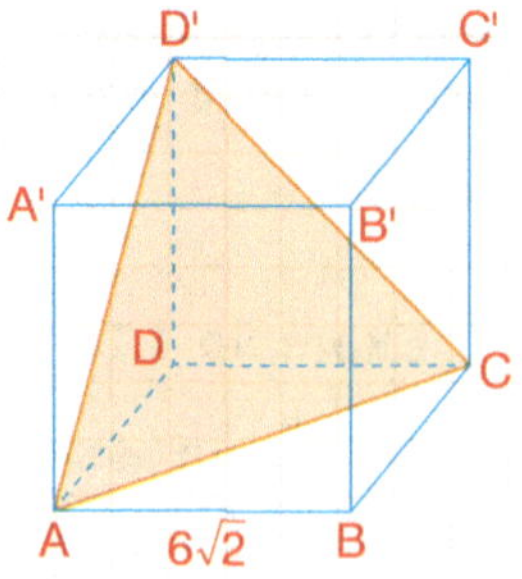

Solution:

As we can see in the figure, $\Delta ACD'$ is an equilateral triangle and we have

$$|AC| = |CD'| = |AD'| = 6\sqrt{2} \cdot \sqrt{2} = 12 \text{ cm}.$$

The area of this triangle is therefore

$$\frac{AB + CD}{BC} = \frac{|AC|^2 \cdot \sqrt{3}}{4} = \frac{12^2 \cdot \sqrt{3}}{4} = 36\sqrt{3} \text{ cm}^2.$$

Example 1.4

Find the total surface area of rectangular prism whose dimensions are in the ratio 1 : 2 : 5, and whose space diagonal measures $\sqrt{60}$ cm.

Solution:

Let $a = t$, $b = 2t$, $c = 5t$.

$$\sqrt{t^2 + (2t)^2 + (5t)^2} = \sqrt{30t^2} = \sqrt{60} \Rightarrow t = \sqrt{2}.$$

Then $a = \sqrt{2}$, $b = 2\sqrt{2}$, $c = 5\sqrt{2}$.

$$A = 2(ab + bc + ac) = 2(4 + 20 + 10) = 68 \text{ cm}^2.$$

Property 1.2

- The **total surface area** of any prism is the sum of the areas of its two bases and its lateral faces:

$$A = 2A_{base} + A_{lateral}$$

- For a **right prism**:

$$A = 2A_{base} + hP_{base}$$

For an **oblique prism**:

$$A = 2A_{base} + lP_{right\ section}$$

Example 1.5

A rectangular prism has edge lengths x, y, z. The areas of its three pairwise-adjacent faces are

$$xz = 24, \quad xy = 32, \quad yz = 36.$$

Find the volume of the prism.

Solution:

Multiply the three equations:

$$(xy)\,(yz)\,(xz) = x^2\, y^2\, z^2 = 24 \cdot 32 \cdot 36.$$

Hence

$$xyz = \sqrt{24 \cdot 32 \cdot 36} = \sqrt{(2^3 \cdot 3)\,(2^5)\,(2^2 \cdot 3^2)} = \sqrt{2^{10} \cdot 3^3}$$

$$= 2^5 \cdot 3^{3/2} = 96\sqrt{3}.$$

So the **volume** is

$$96\sqrt{3}.$$

(Optionally, the edges are

$$x = \sqrt{\frac{24 \cdot 32}{36}} = \frac{8}{\sqrt{3}}, \quad y = \frac{32}{x} = 4\sqrt{3}, \quad z = \frac{24}{x} = 3\sqrt{3},$$

confirming $xyz = 96\sqrt{3}$.)

Theorem 1.2

The distance between two points $O(x_1, y_1, z_1)$ and $Q(x_2, y_2, z_2)$ is

$$PQ = \sqrt{(x_2 - x_1)^2 + (y_2 - y_1)^2 + (z_2 - z_1)^2}.$$

Proof 1.1

In the xy-plane, the distance from $P(x_1, y_1, z_1)$ to $Q(x_2, y_2, z_2)$ is $PQ = \sqrt{(x_2 - x_1)^2 + (y_2 - y_1)^2}$.

Since triangle PRQ is right-angled at R, applying the Pythagorean theorem again gives

$$PQ^2 = (x_2 - x_1)^2 + (y_2 - y_1)^2 + (z_2 - z_1)^2,$$

so

$$PQ = \sqrt{(x_2 - x_1)^2 + (y_2 - y_1)^2 + (z_2 - z_1)^2}.$$

Theorem 1.3

A cube is the three-dimensional version of a square. We know the following information if the cube has side length s:

- The volume is s^3.
- The surface area is $6s^2$.
- The longest distance between two points in the cube is the distance between opposite vertices, known as the space diagonal, and is equal to $s\sqrt{3}$.

Proof 1.2

- **Volume:** $V = s \cdot s \cdot s = s^3$ (length x width x height).
- **Surface area:** 6 square faces, each area $s^2 \Rightarrow A = 6s^2$.
- **Space diagonal:** Coordinates (0, 0, 0) to (s, s, s):

  $d = \sqrt{s^2 + s^2 + s^2} = s\sqrt{3}$.

(Equivalently: face diagonal $s\sqrt{2}$, then $\sqrt{(s\sqrt{2})^2 + s^2} = s\sqrt{3}$.)

Theorem 1.4

The **volume** of any prism is

$$V = Bh,$$

where B is the area of the base and h is the perpendicular height.

If the height is perpendicular to the bases, the **surface area** is

$$A = 2B + h \times (\text{perimeter of base}).$$

Theorem 1.5

The volume of any pyramid is

$$V = \frac{1}{3}Bh,$$

where B is the area of the base and h is the perpendicular height.

Definition 1.5

A sphere is the set of all points in space that are the same distance from a fixed point, called the center. Balls and globes are examples of spheres.

The only key measurement of a sphere is its radius, the distance from the center to any point on the surface.

The formulas for the volume and surface area of a sphere are difficult to derive (Archimedes did so and considered it his greatest discovery).

Formula for a sphere of radius r:

$$\text{Volume} = \frac{4}{3}\pi r^3, \quad \text{Surface Area} = 4\pi r^2$$

Example 1.6

A plane cuts through the interior of a sphere. Let the center of the sphere be O. From O, drop a perpendicular to the plane, meeting it at X. Let A be any point where the plane meets the surface of the sphere.

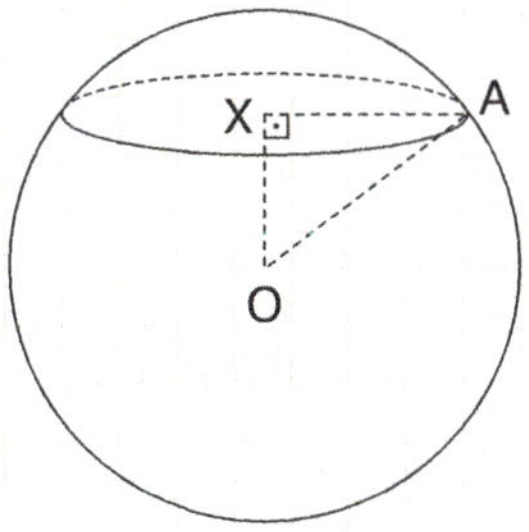

(a) What kind of triangle is ΔOAX?

(b) Express XA in terms of OA and OX.

MathTopia Press

Solution:

Let the sphere have center O, radius $r = OA$. Let the cutting plane be at perpendicular distance $OX = d$ from O. Take any intersection point A in the plane.

(a) Since OX is perpendicular to the plane, it is perpendicular to XA.

Thus $\angle OXA = 90°$ and ΔOAX is **right** at X.

(b) By Pythagoras in ΔOAX,

$$OA^2 = OX^2 + XA^2 \Rightarrow XA = \sqrt{OA^2 - OX^2}.$$

Hence all such points A are at the same distance XA from X; the intersection is a **circle** with center X and radius

$$\sqrt{OA^2 - OX^2}.$$

Example 1.7

A pyramid has a square base ABCD of area 196, and vertex E. The areas of triangles ABE and CDE are 105 and 91. Find the volume of the pyramid.

Solution:

Side of base $AB = 14$.

From $\frac{1}{2}(14)(PE) = 105 \Rightarrow PE = 15$, and

$\frac{1}{2}(14)(QE) = 91 \Rightarrow QE = 13$.

Triangle EPQ is a 13 – 14 – 15 right triangle, so its height (from E to base ABCD) is 12. Thus,

$$V = \frac{1}{3}Bh = \frac{1}{3}(196)(12) = 784.$$

Theorem 1.5

A **cylinder** is a prism with a **circular base** of radius r and height h. Then:

- **Volume:** $V = \pi r^2 h$
- **Surface Area:** $A = 2\pi r^2 + 2\pi rh$

Proof 1.3

The volume of any prism equals the area of its base times its height.

Since the base of a cylinder is a circle with area πr^2, the volume is $\pi r^2 h$.

The surface area includes two circular bases ($2\pi r^2$) and one rectangular lateral face ($2\pi rh$) formed by unrolling the curved surface.

Theorem 1.6

A **cone** is a pyramid with a **circular base** of radius r and height h.

- **Volume:** $V = \frac{1}{3}\pi r^2 h$
- **Surface Area:** $A = \pi r^2 + \pi r\sqrt{r^2 + h^2}$

  or equivalently $A = \pi r^2 + \pi r l$, where $l = \sqrt{r^2 + h^2}$ is the **slant height**.

Proof 1.4

The volume $V = \frac{1}{3}\pi r^2 h$ follows from the pyramid volume rule.

For the surface area, the base contributes πr^2.

To find the lateral area, imagine cutting and unrolling the cone's surface – it forms a **sector** of a circle with radius l.

If it were a full circle, the area would be πl^2, but the cone's sector covers only a fraction equal to the ratio of the base's circumference to the full circle's circumference:

$$\frac{2\pi r}{2\pi l} = \frac{r}{l}.$$

Thus, the lateral area is $\pi l^2 \cdot \frac{r}{l} = \pi r l$.

Example 1.8

A right circular cone has height 9 and base radius 12. A plane perpendicular to the axis of the cone cuts the cone at a point on the axis that is 6 units from the vertex, removing a smaller cone from the top.

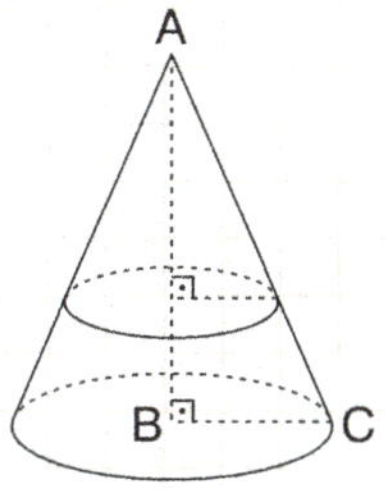

(a) What is the height and volume of the small (top) cone?

(b) What is the volume of the remaining piece (a frustum)?

(c) What is the ratio of the volume of the small cone to the volume of the original cone?

Solution:

The cut is parallel to the base, so the top piece is a cone similar to the original with linear scale factor

$$k = \frac{6}{9} = \frac{2}{3}.$$

(a)

Height of small cone: $h_s = 6$.

Radius of small cone: $r_s = k \cdot 12 = \frac{2}{3} \cdot 12 = 8$.

Volume:

$$V_s = \frac{1}{3}\pi r_s^2 h_s = \frac{1}{3}\pi(8)^2(6) = \frac{1}{3}\pi \cdot 64 \cdot 6 = 128\pi.$$

(b)

Volume of original cone:

$$V = \frac{1}{3}\pi(12)^2(9) = \frac{1}{3}\pi \cdot 144 \cdot 9 = 432\pi.$$

Frustum volume (the other piece):

$$V_f = V - V_s = 432\,\pi - 128\,\pi = 304\,\pi.$$

(c)

Ratio small to original:

$$\frac{V_s}{V} = \frac{128\pi}{432\pi} = \frac{8}{27} = (\frac{2}{3})^3.$$

Example 1.9

In an **oblique** circular cylinder, the top and bottom bases have centers A and B, respectively. A point C lies on the **circumference** of the bottom base. You are given

$AB = 8$, $BC = 4$, $\angle ACB = 90°$.

Find the **height** of the cylinder (the perpendicular distance between the two bases).

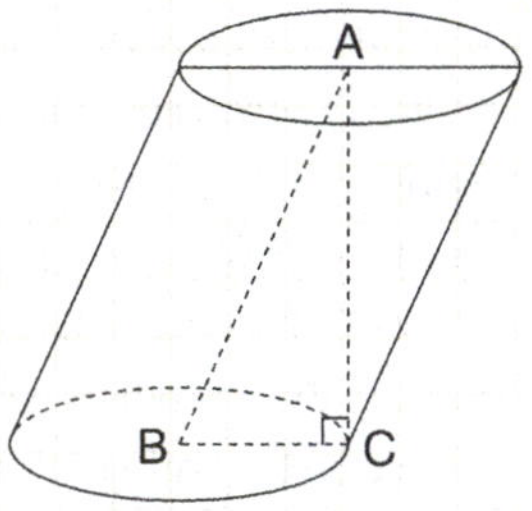

Solution:

Triangle ΔACB is right at C. Since B is the center of the bottom base, BC is a radius of length 4. Segment AC is drawn **perpendicular to the base** through C (so it is an altitude from the top base to the bottom base), and thus its length equals the cylinder's height h.

By the Pythagorean Theorem in right triangle ΔACB with hypotenuse $AB = 8$:

$$AC^2 + BC^2 = AB^2 \Rightarrow h^2 + 4^2 = 8^2 \Rightarrow h2 = 64 - 16 = 48.$$

Hence

$$h = AC = 4\sqrt{3}.$$

Example 1.10

Find the total surface area of an isosceles trapezoidal prism with height 12 cm if the parallel edges of the base are 2 cm and 8 cm and the legs of the base are 5 cm each.

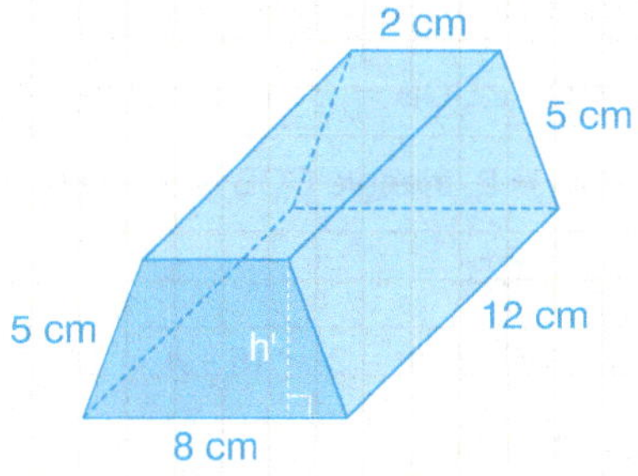

Solution:

The perimeter of the base is $P_{base} = 2 + 8 + 5 + 5 = 20$ cm

Since the base is an isosceles trapezoid, $h' = 4$ cm and so

$$A_{base} = \frac{1}{2}h \cdot (b1 + b2) = \frac{1}{2}4 \cdot (2 + 8) = 20 \text{ cm}^2.$$

Therefore the total area is

$$A_S = 2 \cdot A_{base} + P_{base} \cdot h = 2 \cdot 20 + 20 \cdot 12 = 280 \text{ cm}^2.$$

Example 1.7

The diagram shows a circle with centre O which lies in a horizontal plane. The diameter AB has length 4. Point P lies vertically above O and PO = $2\sqrt{2}$. Point C lies on the semicircular arc AB such that the ratio of the lengths of the arcs AC and CB is 2 : 1.

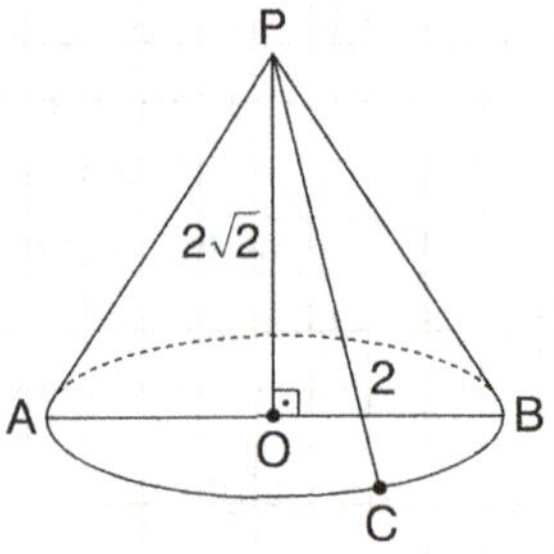

What is the shortest distance from A to PC?

Solution:

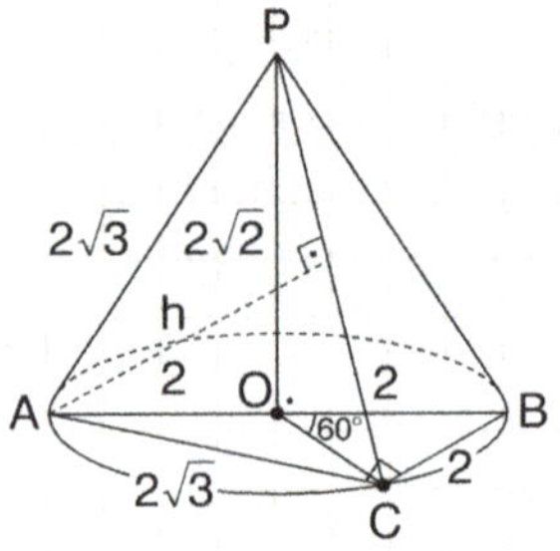

Let the circle have centre O and diameter AB = 4, so AO = BO = 2.

Point P is vertically above O such that PO = $2\sqrt{2}$.

Step 1. Find PA

In right triangle POA,

$PA^2 = AO^2 + PO^2 = 2^2 + (2\sqrt{2})^2 = 4 + 8 = 12 \Rightarrow PA = 2\sqrt{3}$.

Step 2. Locate C

Given that the arcs AC : CB = 2 : 1,

$$\angle AOC : \angle COB = 2 : 1 \Rightarrow \angle COB = 60°.$$

Since OB = OC = 2, triangle COB is equilateral, so BC = 2.

In right triangle ACB,

$$AC^2 = AB^2 - BC^2 = 4^2 - 2^2 = 16 - 4 = 12 \Rightarrow AC = 23.$$

Step 3. Triangle PAC

Triangle POA and ACB give PA = AC = $2\sqrt{3}$.

Also, PC = $2\sqrt{3}$. Hence, ΔPAC is equilateral with side $2\sqrt{3}$.

Step 4. Shortest distance from A to PC

This is the height h of the equilateral triangle:

$$h = \frac{\sqrt{3}}{2} \times 2\sqrt{3} = 3.$$

Practice 1

Let M be the midpoint of $\overline{AB}$ in regular tetrahedron ABCD. What is cos(∠CMD)?

(A) $\frac{1}{4}$ (B) $\frac{1}{3}$ (C) $\frac{2}{5}$ (D) $\frac{1}{2}$ (E) $\frac{\sqrt{3}}{2}$

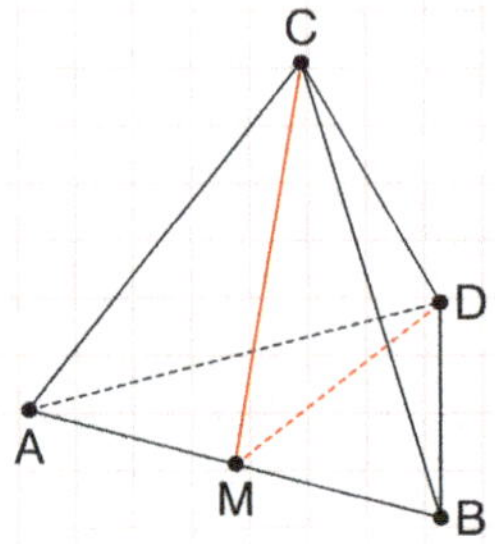

Solution:

Without loss of generality, let the edge-length of ABCD be 2. It follows that MC = MD = $\sqrt{3}$.

Let O be the center of ΔABD, so $\overline{CO} \perp \overline{MD}$. Note that

$$MO = \frac{1}{3}MD = \frac{\sqrt{3}}{3}.$$

In right ΔCMO, we have $\cos(\angle CMD) = \frac{MO}{MC} = \frac{1}{3}$.

Practice 2

Join the centers of adjacent faces of a unit cube to form a regular octahedron. Find its volume.

Solution:

Distance between centers of two adjacent faces of a unit cube:

$$s = \sqrt{\left(\frac{1}{2}\right)^2 + \left(\frac{1}{2}\right)^2} = \frac{1}{\sqrt{2}}.$$

A regular octahedron with side s has volume

$$V = \frac{\sqrt{2}}{3}s^3 = \frac{\sqrt{2}}{3}\left(\frac{1}{\sqrt{2}}\right)^3 = \frac{\sqrt{2}}{3} \cdot \frac{\sqrt{2}}{4} = \frac{2}{12} = \frac{1}{6}.$$

Remark (alternate view): The octahedron is two congruent square pyramids with base side $s\sqrt{2} = 1$ and height $\frac{1}{2}$, so

$$V = 2 \cdot \frac{1}{3}(1^2)\left(\frac{1}{2}\right) = \frac{1}{6}.$$

1. Triangle PAB and square ABCD are in perpendicular planes at right. Given that PA = 3, PB = 4, and AB = 5, what is PD?

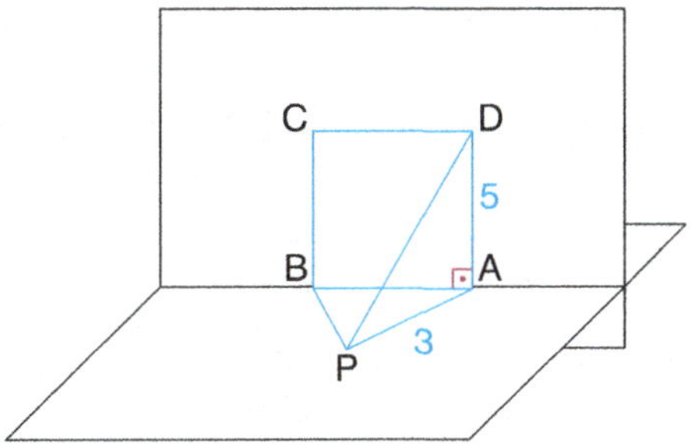

2. Bridget places a box that is 4 x 6 x 8 on the floor. She then places a 2 x 3 x 5 box on top of the first box, forming a two-box tower. She will then paint all the surfaces she can paint without moving either of the boxes. She wants to paint as little as possible, so she places the boxes in a way that minimizes the amount she'll have to paint. What is the total area she has to paint?

Type your solution, notes and/or work here.

3. A cylinder is inscribed in a square prism. As shown at left, each of the bases of the cylinder is inscribed in a face of the prism, and the cylinder is tangent to the other four faces of the prism. Given that AB = 4 and AE = 8, find the volume of the cylinder.

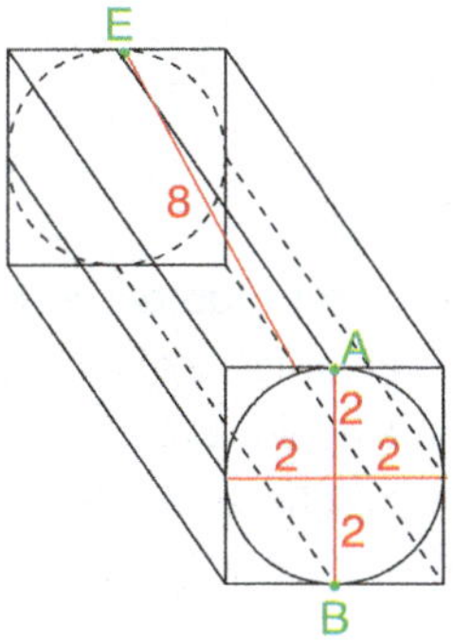

4. For the adjacent square right frustum, find the slant height of the darker lateral face.

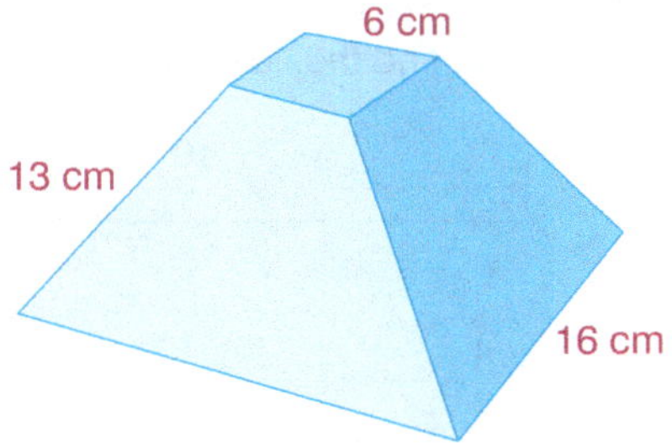

5. A plane cuts through a 26 cm diameter sphere, but the closest it gets to the center is 5 cm. What is the area of the intersection of the sphere and the plane?

6. A right hexagonal prism has bases that are regular hexagons with side length 5 cm and height 12 cm. Only the six rectangular lateral faces are painted at a rate of 30 cents per square centimeter. What is the total cost?

7. A cylindrical bottle containing some water is lying horizontally on the ground. If the radius of the base is 6 cm, the length of the bottle is 10 cm and the surface of the water is 3 cm from the ground, find the volume of water in the bottle.

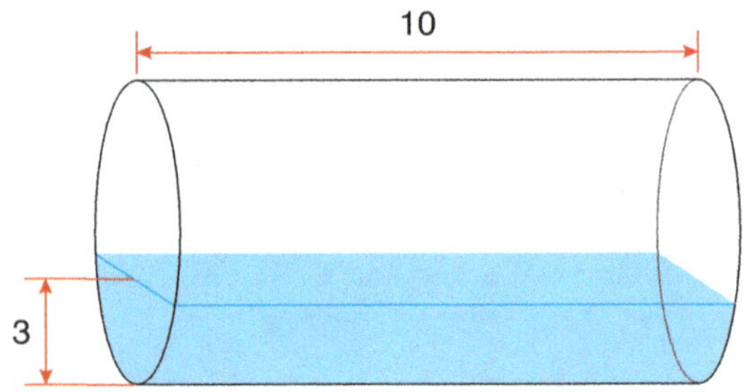

8. A cuboid has sides of lengths 22, 2 and 10. It is contained within a sphere of the smallest possible radius.

What is the side-length of the largest cube that will fit inside the same sphere?

A) 10 B) 11 C) 12 D) 13 E) 14

9. In a right circular cone with base radius 6 and height 12, a largest possible sphere is inscribed (it is tangent to the base and the lateral surface). What is the surface area of this sphere?

10. **(08-2000-022) (AMC8-2000-22)**

A cube has edge length 2. Suppose that we glue a cube of edge length 1 on top of the big cube so that one of its faces rests entirely on the top face of the larger cube. The percent increase in the surface area (sides, top, and bottom) from the original cube to the new solid formed is closest to:

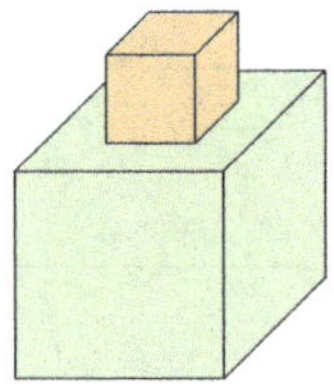

A) 10 B) 15 C) 17 D) 21 E) 25

11. A regular tetrahedron is a polyhedron with four faces, each of which is an equilateral triangle. A solid regular tetrahedron is cut into two pieces by a single plane cut.

Which of the following could not be the shape of the section formed by the cut?

A) a pentagon

B) a square

C) a rectangle that is not a square

D) a trapezium

E) a triangle that is not equilateral

12. **(UNB-2007-Gr 9-18)**

60 cubes are glued together in a 3 x 4 x 5 rectangular shape. The entire outside of the object is painted. Then, it is broken apart again, back to the original cubes. How many of the cubes have exactly one painted face?

A) 11 B) 22 C) 33 D) 44 E) 52

1. What is the volume of this geometric figure?

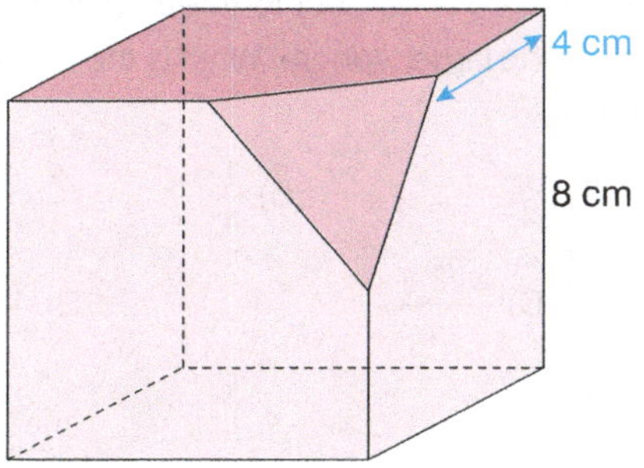

2. Three unit cubes are glued together so that any two touching cubes share a whole face (the solid is face-connected). Which of the following cannot be the distance between two vertices of the resulting solid?

A) $\sqrt{7}$ B) $\sqrt{8}$ C) $\sqrt{9}$ D) $\sqrt{10}$ E) $\sqrt{11}$

3. A right square prism has three faces painted white and the other three faces painted red. The sum of the areas of the white faces is 76 square units, and the sum of the areas of the red faces is 12 square units.

What is the volume of the prism (in cubic units)?

A) 18 B) 24 C) 27 D) 32 E) 36

4. **(UNB-2007-Gr 9-18)**

A solid right, circular cone has a base radius of 5 meters and slant height of 10 meters. A top cone will be removed so that the volume of the remaining frustum will be 1/3 that of the original cone. In square meters, what is the total surface area, including both bases, of the frustum? Express your answer to the nearest whole number.

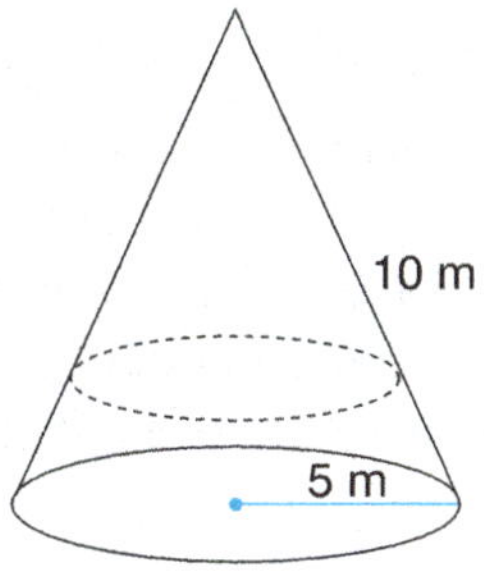

5. **(TÜBİTAK 1999, High School Math Olympiad):**

A sphere is inscribed in a right truncated cone whose base radii are 2 and 6. The sphere is tangent to both bases and to the lateral surface. What is the volume of this truncated cone?

A) $\frac{148\sqrt{3}}{2}\pi$ B) $\frac{164\sqrt{3}}{3}\pi$ C) $\frac{208\sqrt{3}}{2}\pi$

D) $\frac{248\sqrt{3}}{2}\pi$ E) $\frac{324\sqrt{3}}{2}\pi$

6. The base ABC is an equilateral triangle and the apex is point T of a regular pyramid. The midpoints of the edges [AB], [BC], [CT], [TA] are denoted by P, Q, R, S, respectively. Given that the height of this pyramid is $2\sqrt{15}$ and $AB = 6$, what is the area of quadrilateral PQRS?

7.

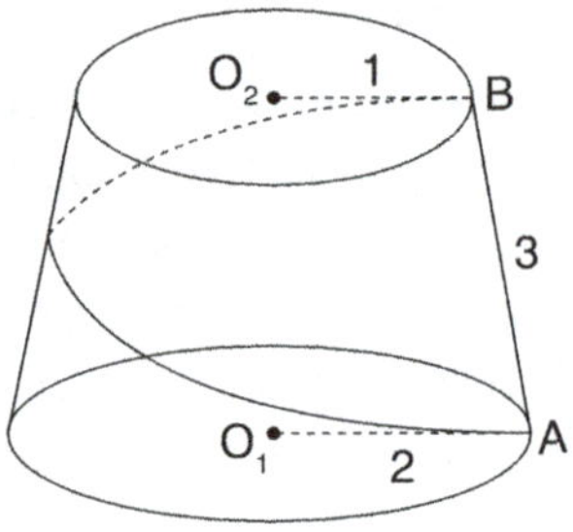

A frustum of a right circular cone has a lower base radius of 2 meters, an upper base radius 1 meter, and slant height $|AB| = 3$ meters. An ant starts from point A on the lower base, makes exactly one complete turn around the frustum along the lateral surface, and arrives at point B on the upper base. What is the length of the shortest path the ant could have taken?

A) $2\sqrt{3}$ B) $3\sqrt{2}$ C) $3\sqrt{3}$ D) $4\sqrt{3}$ E) None of these

8. **(AMC8-2008-21)**

Jerry cuts a wedge from a 6 cm cylinder of bologna as shown by the dashed curve. What is the volume of his wedge in cubic centimeters?

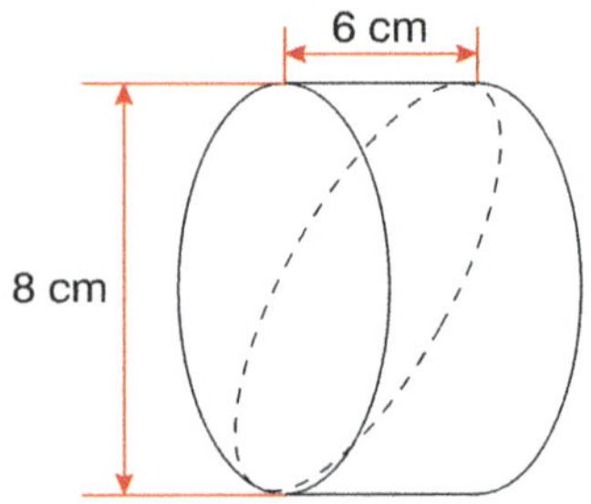

9. The right hexagonal prism shown below has all edges of length 1. An ant crawls from point A to point B along the surface of the hexagonal prism. What is the shortest possible distance the ant can crawl?

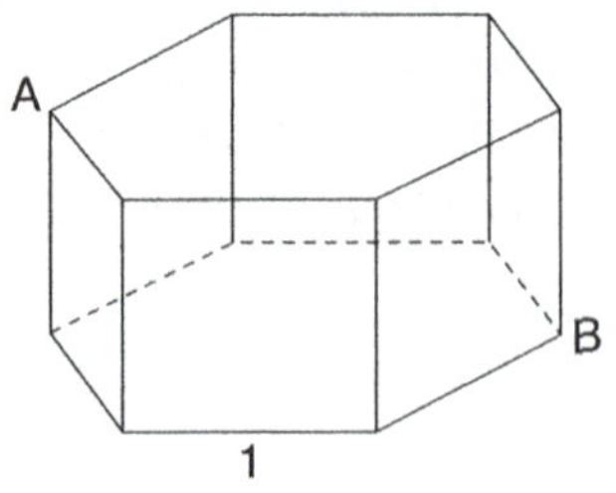

A) $\sqrt{5 + \sqrt{3}}$ B) $\sqrt{5 + 2\sqrt{3}}$ C) $3\sqrt{3}$ D) 3 E) $\sqrt{2} + \sqrt{3}$

10. A sphere is inscribed inside a cube of side length 2 cm. A smaller sphere is tangent to three of the faces of the cube as well as the larger sphere. What is the radius of the smaller sphere?

A) $5 - 3\sqrt{3}$ B) $\frac{1}{4}$ C) $2 - \sqrt{3}$ D) $\frac{3\sqrt{3} - 4}{4}$ E) $\frac{\sqrt{3} - 1}{2}$

11. **(UKMT, Senior Mathematics Challenge 2011, #16)**

A solid sculpture consists of a 4 x 4 x 4 cube with a 3 x 3 x 3 cube sticking out as shown. Three vertices of the smaller cube lie on edges of the larger cube, the same distance along each edge. What is the total volume of the sculpture?

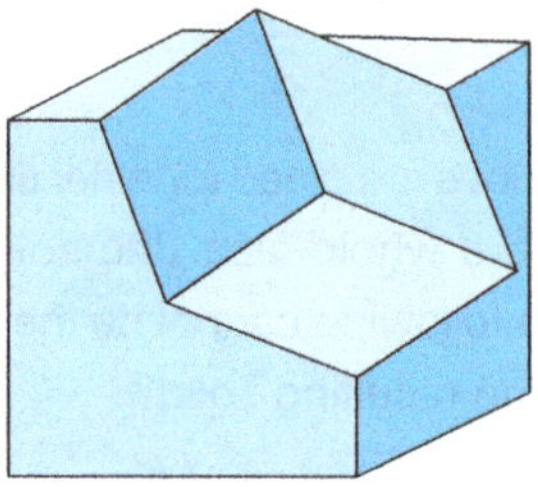

12. A cube has side length 4 inches. A solid is formed by slicing eight congruent tetrahedra from the cube as shown. What is the surface area of the solid, in square inches?

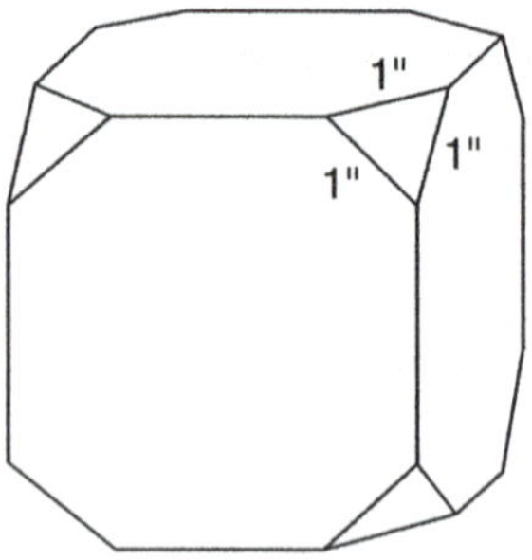

A) $84 + 2\sqrt{3}$ B) $90 + 2\sqrt{3}$ C) $90 + 4\sqrt{3}$ D) $102 - 2\sqrt{3}$ E) None of the preceding

1. **(MathCON 2021 Final GR 11, #10)**

A semicircle paper with center O on the left is folded to get the right cone on the right. Points A_1 and A_2 coincide and O is the vertex of the cone.

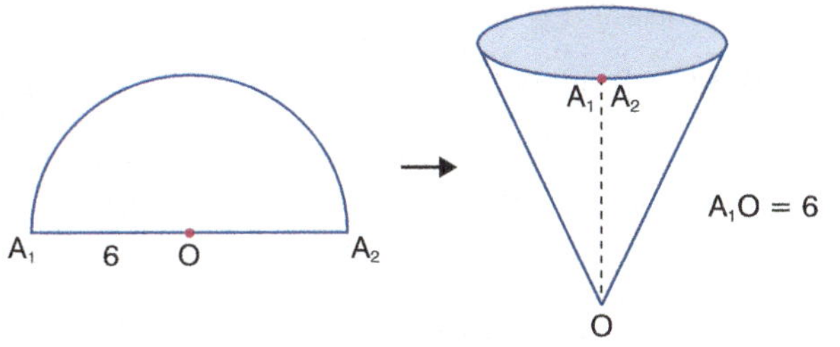

What is the base area of the cone?

A) 6π cm^2 B) 7π cm^2 C) 8π cm^2 D) 9π cm^2 E) 10π cm^2

2. **(MathCON 2021 Final GR 10, #18)**

In the figure, a semi-sphere, a right cylinder, and a right circular cone are glued to each other. The centers of bases of the figures coincide and each of the radius of three of the figures are the same, where O is the center and P is the vertex of the cone.

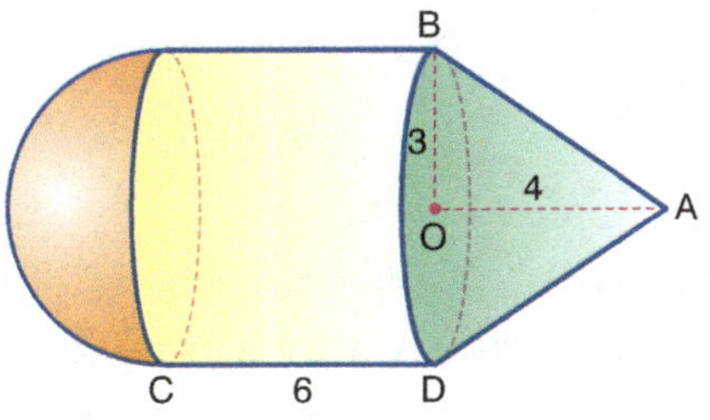

OA = 4 cm,
OB = 3 cm, and
CD = 6 cm.

What is the volume of the figure?

A) 84π cm^3 B) 82π cm^3 C) 80π cm^3 D) 78π cm^3 E) 72π cm^3

3. **(2021 AMC 10B, #10)**

An inverted cone with base radius 12 cm and height 18 cm is full of water. The water is poured into a tall cylinder whose horizontal base has a radius of 24 cm.

What is the height in centimeters of the water in the cylinder?

A) 1.5 B) 3 C) 4 D) 4.5 E) 6

4. **(2020 AMC 10B, #10)**

A three-quarter sector of a circle of radius 4 inches along with its interior is the 2-D net that forms the lateral surface area of a right circular cone by taping together along the two radii shown. What is the volume of the cone in cubic inches?

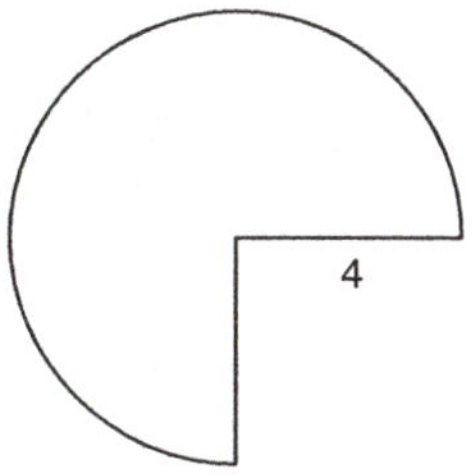

A) $3\pi\sqrt{5}$ B) $4\pi\sqrt{3}$ C) $3\pi\sqrt{7}$ D) $6\pi\sqrt{3}$ E) $6\pi\sqrt{7}$

5. **(2018 AMC 10B, #10)**

In the rectangular parallelepiped shown, AB = 3, BC = 1, and CG = 2. Point M is the midpoint of $\overline{FG}$.

What is the volume of the rectangular pyramid with base BCHE and apex M?

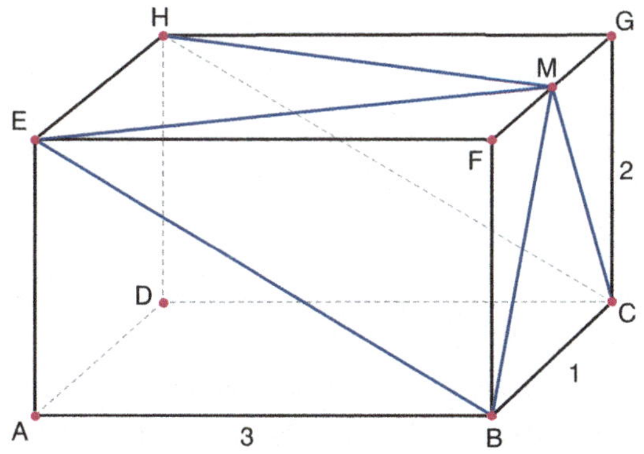

A) 1 B) $\frac{4}{3}$ C) $\frac{3}{2}$ D) $\frac{5}{3}$ E) 2

6. **(2021 AMC 12A, #10)**

Two right circular cones with vertices facing down as shown in the figure below contains the same amount of liquid. The radii of the tops of the liquid surfaces are 3 cm and 6 cm. Into each cone is dropped a spherical marble of radius 1 cm, which sinks to the bottom and is completely submerged without spilling any liquid. What is the ratio of the rise of the liquid level in the narrow cone to the rise of the liquid level in the wide cone?

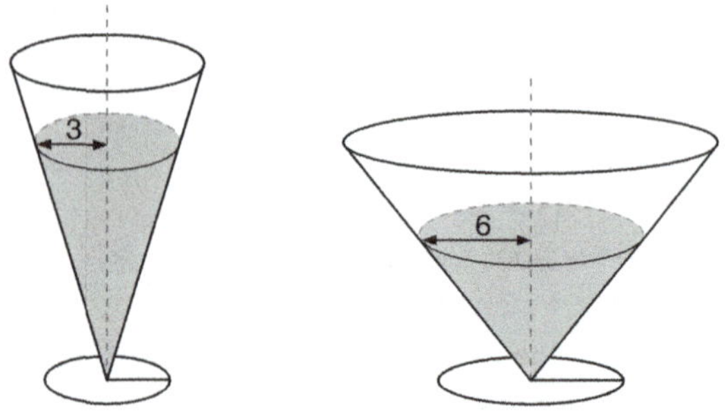

A) 1 : 1 B) 47 : 43 C) 2 : 1 D) 40 : 13 E) 4 : 1

7. **(2017 AMC 10A, #11)**

The region consisting of all points in three-dimensional space within 3 units of line segment $\overline{AB}$ has volume 216π. What is the length AB?

A) 6 B) 12 C) 18 D) 20 E) 24

8. **(2021 AMC 10A, #13)**

What is the volume of tetrahedron ABCD with edge lengths $AB = 2$, $AC = 3$, $AD = 4$, $BC = \sqrt{13}$, $BD = 2\sqrt{5}$, and $CD = 5$?

A) 3 B) $2\sqrt{3}$ C) 4 D) $3\sqrt{3}$ E) 6

9. **(2014 AMC 12B, #14)**

A rectangular box has a total surface area of 94 square inches. The sum of the lengths of all its edges is 48 inches. What is the sum of the lengths in inches of all of its interior diagonals?

A) $8\sqrt{3}$ B) $10\sqrt{2}$ C) $16\sqrt{3}$ D) $20\sqrt{2}$ E) $40\sqrt{2}$

10. **(2004 AMC 10A, #19)**

A white cylindrical silo has a diameter of 30 feet and a height of 80 feet. A red stripe with a horizontal width of 3 feet is painted on the silo, as shown, making two complete revolutions around it. What is the area of the stripe in square feet?

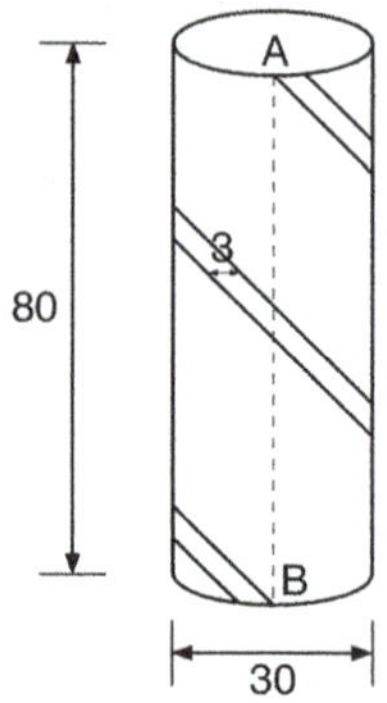

A) 120 B) 180 C) 240 D) 360 E) 480

11. **(2001 AMC 10, #21)**

A right circular cylinder with its diameter equal to its height is inscribed in a right circular cone. The cone has diameter 10 and altitude 12, and the axes of the cylinder and cone coincide. Find the radius of the cylinder.

A) $\frac{8}{3}$ B) $\frac{30}{11}$ C) 3 D) $\frac{25}{8}$ E) $\frac{7}{2}$

12. **(2015 AMC 10A, #21)**

Tetrahedron ABCD has $AB = 5$, $AC = 3$, $BC = 4$, $BD = 4$, $AD = 3$, and $CD = \frac{12}{5}\sqrt{2}$.

What is the volume of the tetrahedron?

A) $3\sqrt{2}$ B) $2\sqrt{5}$ C) $\frac{24}{5}$ D) $3\sqrt{3}$ E) $\frac{24}{5}\sqrt{2}$

CHAPTER 8

Out-of-the-Box Geometry: AIME Connections

Out-of-the-Box Geometry: AIME Connections

1- Geometric Reasoning and Visual Inequalities

2- Trigonometric & Circle Applications

3- Advanced Formulas& Theorems

The Inequality

For nonnegative real numbers $x_1, x_2, ..., x_n$,

$$\frac{x_1 + x_2 + ... + x_n}{n} \geq \sqrt[n]{x_1 x_2 ... x_n},$$

with equality when $x_1 = x_2 = ... = x_n$.

This is the arithmetic mean-geometric mean inequality (AM-GM).

It's powerful tool in AMC 10-12 geometry problems, often used to maximize areas or minimize lengths.

Example 1.1

Rectangle with fixed perimeter

A rectangle has perimeter 40. Find its maximum area.

Solution:

- Sides: $x + y = 20$.
- Area $A = xy$.
- By AM-GM: $\frac{x+y}{2} \geq \sqrt{xy} \Rightarrow 10 \geq \sqrt{A}$.
- So $A \leq 100$.

Maximum area 100, attained when $x = y = 10$ (a square).

Example 1.2

Triangle with fixed perimeter

Among all triangles with perimeter P, which has the greatest area?

Solution:

- By Heron's formula, $K = \sqrt{s(s-a)(s-b)(s-c)}$ with $s = \frac{P}{2}$.
- By AM-GM, $(s - a)\ (s - b)\ (s - c) \leq \left(\frac{s}{3}\right)^3$.
- So $K \leq \frac{s^2}{3\sqrt{3}} = \frac{P^2}{12\sqrt{3}}$.

Equality occurs when $a = b = c$. The equilateral triangle has max area.

Example 1.3

Triangle with fixed base + height

A triangle has base b and height h with $b + h = 20$.

Solution:

- Area: $K = \frac{1}{2}bh$.
- By AM-GM: $bh \leq \left(\frac{b+h}{2}\right)^2 = 100$
- So $K \leq 50$.

Maximum area 50 when $b = h = 10$.

Example 1.4

In ΔABC with $AB = 3$, $BC = 4$, $CA = 5$, points $M \in \overline{BC}$ and $N \in \overline{AC}$ are chosen so that segment $\overline{MN}$ divides ΔABC into two regions of equal area. What is the minimum possible length of $\overline{MN}$?

Solution:

ΔABC is 3-4-5 triangle with right angle at B.

Let $M \in \overline{BC}$ and $N \in \overline{AC}$ so that MN cuts the triangle into equal areas.

Let $u = CM$ (along BC) and $v = CN$ (along AC). Angle at C satisfies

$$\cos C = \frac{4}{5}, \quad \sin C = \frac{3}{5}.$$

Equal areas means the small triangles ΔCMN has area 3:

$$\frac{1}{2} u v \sin C = 3 \Rightarrow uv = \frac{6}{\sin C} = 10. \qquad (1)$$

By the law of cosines in ΔCMN,

$$MN^2 + u^2 + v^2 - 2uv \cos C. \qquad (2)$$

The product uv is fixed (by(1)). To minimize MN^2, minimize $u^2 + v^2$ subject to $uv = 10$.

By AM–GM, $u^2 + v^2 \geq 2uv = 20$, with equality at $u = v$.

Plug into (2):

$$MN^2_{min} = 20 - 2(10)\cos C = 20\left(1 - \frac{4}{5}\right) = 4 \Rightarrow MN_{min} = 2.$$

Equality occurs when $u = v = \sqrt{10}$(i.e., $CM = CN = \sqrt{10}$).

2

MathTopia Press

Example 1.5

In convex quadrilateral ABCD, diagonals intersect at O.

If [AOB] = 4 and [COD] = 9, find the minimum possible area of ABCD.

Solution:

Let [AOB] = 4, [COD] = 9, [BOC] = y, [DOA] = w.

Since $\frac{AO}{OC} = \frac{[AOB]}{[BOC]} = \frac{[AOD]}{[COD]} \Rightarrow \frac{4}{y} = \frac{w}{9} = k$,

then $y = \frac{4}{k}$, $w = 9k$.

$$S = 4 + y + 9 + w = 13 + \frac{4}{k} + 9k.$$

Minimize S.

By AM–GM: $\frac{4}{k} + 9k \geq 12 \Rightarrow S_{min} = 13 + 12 = 25$.

Example 1.6

In ΔABC, ∠BAC = 60° and BC = 2. Find the maximum value of $3AB^2 + 8AC^2$.

Solution:

Let AB = c, AC = b, BC = a = 2.

By the cosine rule, $a^2 = b^2 + c^2 - bc \Rightarrow 4 = b^2 + c^2 - bc$.

We need $S = 8b^2 + 3c^2$.

Multiply the given equation by 3:

$$12 = 3b^2 + 3c^2 - 3bc \Rightarrow 8b^2 + 3c^2 = 12 + 5b^2 + 3bc.$$

By

$2xy \leq x^2 + y^2$, we have $4b^2 + 9c^2 \geq 12bc \Rightarrow \leq 4b^2 + 9c^2$

Hence, $8b^2 + 3c^2 \leq 12a^2 = 12(2^2) = 48$.

Equality holds when 2b = 3c.

Example 1.7

What is the maximum area of a quadrilateral whose sides have a length of 23 and the circumference of 62?

Solution:

Among all quadrilaterals with a fixed perimeter, the cyclic quadrilateral has the greatest area.

Using Brahmagupta's Formula:

$S = \sqrt{(s-a)(s-b)(s-c)(s-d)}$ where $s = \frac{a+b+c+d}{2}$.

Given: $d = 23$ and $a + b + c + d = 62 \Rightarrow a + b + c = 39$, $s = 31$. Then, $S^2 = (31 - a)(31 - b)(31 - c)(31 - 23)$.

Let $P = (31 - a)(31 - b)(31 - c)$.

By AM–GM,

$$P \leq \left(\frac{31 - a + 31 - b + 31 - c}{3}\right)^3 = 18^3.$$

Thus,

$$S^2 \leq 8 \times 18^3 = 36^3 \Rightarrow s \leq 216.$$

Maximum area: $S_{max} = 216$

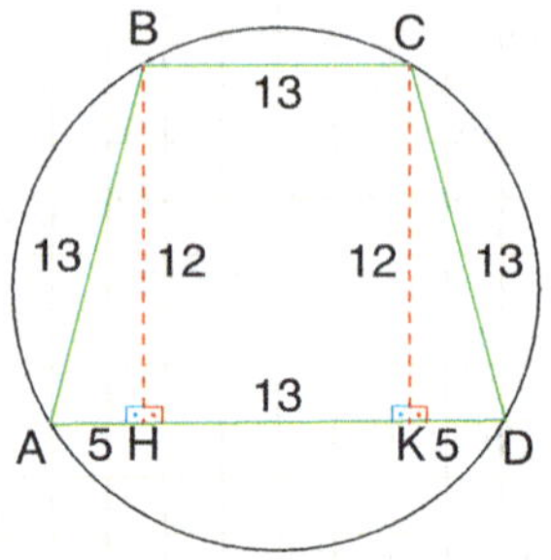

MathTopia Press

Example 1.8

A rectangular block of cheese measures 10 × 13 × 14 cm. Ten 1 cm–thick slices are cut parallel to any faces (not necessarily the same). What is the maximum possible volume of the remaining block?

Solution:

Each slice reduces one dimension by 1.

Initially, a + b + c = 10 + 13 + 14 = 37.

After ten slices, a + b + c = 27.

We want to maximize V = abc given a + b + c = 27.

By AM-GM, $abc \leq \left(\frac{a+b+c}{3}\right)^3 = 9^3 = 729$.

Theorem 1.1

Trigonometric Identities and Transformations

1. Sum and Difference Formulas

$\cos(A + B) = \cos A \cos B - \sin A \sin B$, $\cos(A - B) = \cos A \cos B + \sin A \sin B$,

$\sin(A + B) = \sin A \cos B + \cos A \sin B$, $\sin(A - B) = \sin A \cos B - \cos A \sin B$,

$\tan(A + B) = \dfrac{\tan A + \tan B}{1 - \tan A \tan B}$, $\tan(A - B) = \dfrac{\tan A - \tan B}{1 + \tan A \tan B}$,

$\cot(A + B) = \dfrac{\cot A \cot B - 1}{\cot B + \cot A}$, $\cot(A - B) = \dfrac{\cot A \cot B + 1}{\cot B - \cot A}$,

2. Double and Triple Angle Formulas

$\sin 2A = 2\sin A \cos A$

$\cos 2A = \cos^2 A - \sin^2 A = 2\cos^2 A - 1 = 1 - 2\sin^2 A$

$\tan 2A = \dfrac{2\tan A}{1 - \tan^2 A}$, $\cot 2A = \dfrac{1 - \cos 2A}{1 + \cos 2A}$

$\sin 3A = 3\sin A - 4\sin^3 A$, $\cos 3A = 4\cos^3 A - 3\cos A$,

$\tan 3A = \dfrac{3\tan A - \tan^3 A}{1 - 3\tan^2 A}$

3. Half-Angle Formulas

If $t = \tan\dfrac{A}{2}$, then

$$\sin A = \frac{2t}{1 + t2}, \cos A = \frac{1 - t^2}{1 + t^2}, \tan A = \frac{2t}{1 - t^2}$$

4. Product-to-Sum (Transformation) Formulas

$\cos A + \cos B = 2\cos\dfrac{A+B}{2}\cos\dfrac{A-B}{2}$,

$\cos A - \cos B = y^2\sin\dfrac{A+B}{2}\sin\dfrac{A-B}{2}$,

$\sin A + \sin B = 2\sin\dfrac{A+B}{2}\cos\dfrac{A-B}{2}$,

$\sin A - \sin B = 2\cos\dfrac{A+B}{2}\sin\dfrac{A-B}{2}$.

Theorem 1.2

Trigonometric Ceva's Theorem

In triangle ABC, let point P be an interior point connected to the vertices. Among the six angles formed, the following relationship holds:

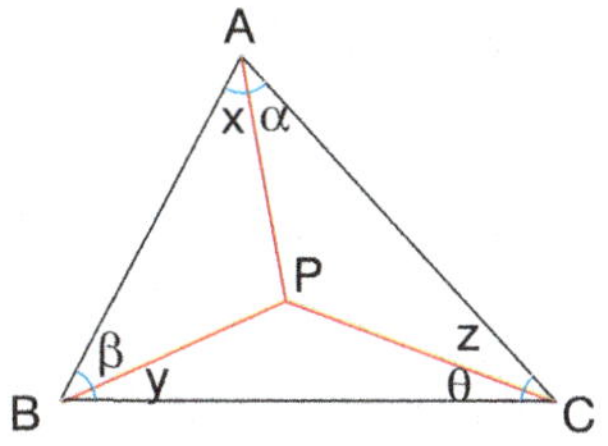

$$\sin x \cdot \sin y \cdot \sin z = \sin \alpha \cdot \sin \beta \sin \theta$$

Example 1.9

In triangle ABC, let

$m(\angle BAC) = 60°$, $|AB| = 8$ cm, $|AC| = 5$ cm.

Inside the triangle, point P satisfies

$m(\angle APB) = m(\angle BPC) = m(\angle APC) = 120°$.

If $|AP| = x$, $|BP| = y$, $|CP| = z$,

find the value of $x + y + z$.

Solution:

Using the cosine rule in triangle ABC: $|BC| = 7$ cm.

Area of ABC: $S = \dfrac{1}{2}(8)(5)\sin 60° = 10\sqrt{3}$

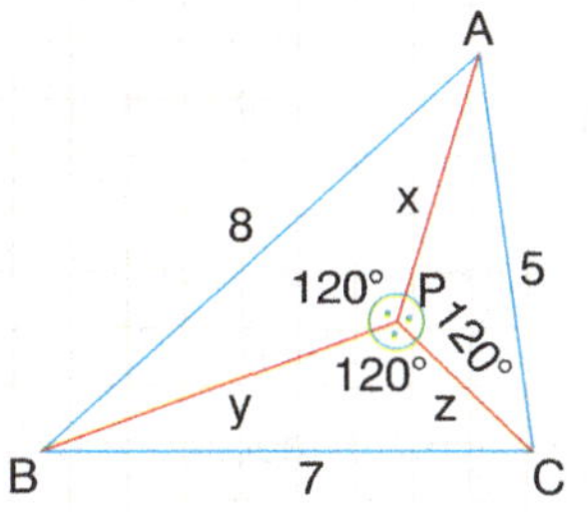

For the three triangles around P:

$$S = \frac{1}{2}(xy + yz + zx)\sin 120° = \frac{\sqrt{3}}{4}(xy + yz + zx).$$

$$\Rightarrow xy + yz + zx = 40.$$

Applying the cosine rule in triangles PAB, PBC, PCA:

$x^2 + y^2 + xy = 64$, $y^2 + z^2 + yz = 49$, $z^2 + x^2 + zx = 25$.

Adding all:

$2(x^2 + y^2 + z^2) + (xy + yz + zx) = 138 \Rightarrow x^2 + y^2 + z^2 = 49$.

Finally

$(x + y + z)^2 + x^2 + y^2 + z^2 + 2(xy + yz + zx) = 49 + 2(40) = 129$. $x + y + z = \sqrt{129}$.

Example 1.10

Let x, y, z be real numbers. What is the maximum value of the expression $\sin x \cos y + \sin y \cos z + \sin z \cos x$?

Solution:

We can use the trigonometric identity

$$\sin A \cos B = \frac{1}{2}[\sin(A + B) + \sin(A - B)].$$

So, $E = \sin x \cos y + \sin y \cos z + \sin z + \cos x = \dfrac{1}{2}[\sin(x + y) + \sin(y + z) + \sin(z + x) + \sin(x - y) + \sin(y - z) + \sin(z - x)]$. Now, note that $\sin(x - y) + \sin(y - z) + \sin(z - x) = 0$.

$$\text{Hence, } E = \frac{1}{2}[\sin(x + y) + \sin(y + z) + \sin(z + x)].$$

To maximize E, all three sine terms should be equal and as large as possible, i.e., $\sin(x + y) = \sin(y + z) = \sin(z + x) = 1$. Thus, $E_{max} = \dfrac{1}{2}(1 + 1 + 1) = \dfrac{3}{2}$.

Example 1.11

In right triangle ABC with $\angle C = 90°$, the hypotenuse BC is divided into three equal parts by points D and E. If $BC = 3AC$, find sum of the angles $\angle AEC + \angle ADC + \angle ABC$.

Solution:

Let $AC = b$. Then $BC = 3b$, so each segment $CD = DE = EB = b$. Denote: $\angle AEC = \alpha$, $\angle ADC = \theta$, $\angle ABC = \beta$.

From the right triangles ADC, ACE, and ACB:

$$\tan\theta = \frac{CD}{AC} = \frac{b}{b} = 1 \Rightarrow \theta = 45°,$$

$$\tan\alpha = \frac{CE}{AC} = \frac{b}{2b} = \frac{1}{2} \Rightarrow \alpha = \arctan\frac{1}{2},$$

$$\tan\beta = \frac{CB}{AC} = \frac{b}{3b} = \frac{1}{3} \Rightarrow \beta = \arctan\frac{1}{3}.$$

Hence, the required sum is:

$$\angle AEC + \angle ADC + \angle ABC = 45° + \arctan\frac{1}{2} + \arctan\frac{1}{3}.$$

Since $\arctan\frac{1}{2} + \arctan\frac{1}{3} = \arctan(1) = 45°$,

$$45° + 45° = 90°$$

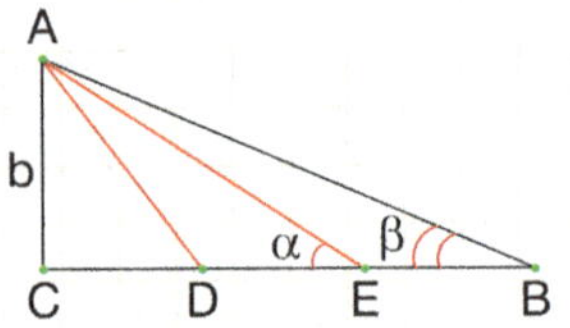

Theorem 1.3

In triangle ABC, if the side lengths are a, b, c and the circumradius R, then

$$A(\triangle ABC) = \frac{a \cdot b \cdot c}{4R}.$$

Theorem 1.4

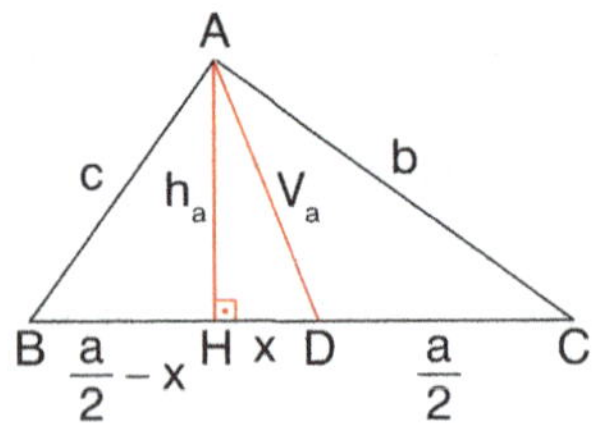

In triangle ABC, let [AH] be the altitude and [AD] the median.

If $|HD| = x$, then the following hold:

a. $V_a^2 = \dfrac{b^2 + c^2}{2} - \dfrac{a^2}{4}$, $\quad V_b^2 = \dfrac{a^2 + c^2}{2} - \dfrac{b^2}{4}$,

$V_c^2 = \dfrac{a^2 + b^2}{2} - \dfrac{c^2}{4}$,

b. $|b^2 - c^2|\ 2ax$

Proof 1.4

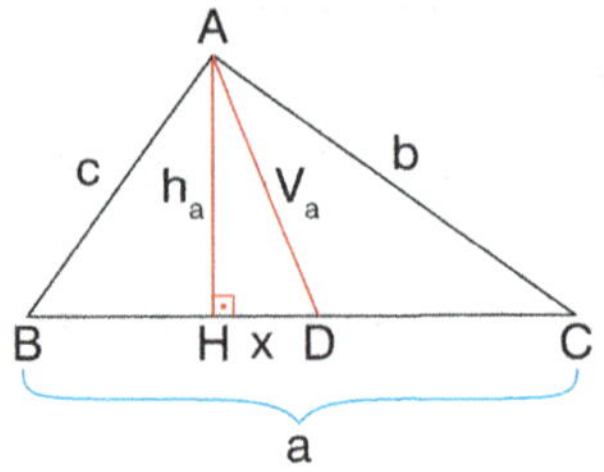

In ΔAHC:

$$h_a^2 + \left(x + \frac{a}{2}\right)^2 = b^2 \qquad (1)$$

In ΔAHB:

$$h_a^2 + \left(\frac{a}{2} - x\right)^2 = c^2 \qquad (2)$$

Also, AHD: $h_a^2 + x^2 = V_a^2$ (3)

Adding (1) and (2): $2h_a^2 + 2x^2 + \dfrac{a^2}{2} = b^2 + c^2$.

Substitute from (3): $2V_a^2 + \dfrac{a^2}{2} = b^2 + c^2 \Rightarrow V_a^2 = \dfrac{b^2 + c^2}{2} - \dfrac{a^2}{4}$.

Similarly, using altitudes h_b, h_c and medians V_b, V_c, we get:

$$V_b^2 = \frac{a^2 + c^2}{2} - \frac{b^2}{4}, \quad V_c^2 = \frac{a^2 + b^2}{2} - \frac{c^2}{4}.$$

Finally, subtracting (1) and (2):

$$b^1 - c^2 = 2ax.$$

Example 1.12

In triangle ABC, let G be the centroid. Show that

$|GA|^2 + |GB|^2 + |GC|^2 = \frac{1}{3}(a^2 + b^2 + c^2)$

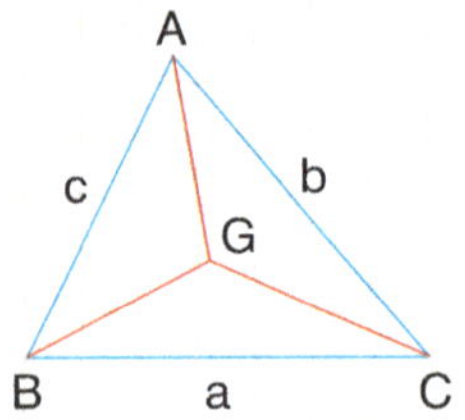

Solution:

Let $|AD| = V_a$, $|BF| = V_b$, $|CE| = V_c$.

Since the centroid divides each median in a 2 : 1 ratio,

$$|GA| = \frac{2}{3}V_a, |GB| = \frac{2}{3}V_b, |GC| = \frac{2}{3}V_c.$$

Squaring and adding:

$$9\left(|GA|^2 + |GB|^2 + |GC|^2\right) = 4\left(V_a^2 + V_b^2 + V_c^2\right).$$

But $V_a^2 + V_b^2 + V_c^2 = \frac{3}{4}(a^2 + b^2 + c^2)$,

so $|GA|^2 + |GB|^2 + |GC|^2 =$

$\frac{1}{3}(a^2 + b^2 + c^2)$

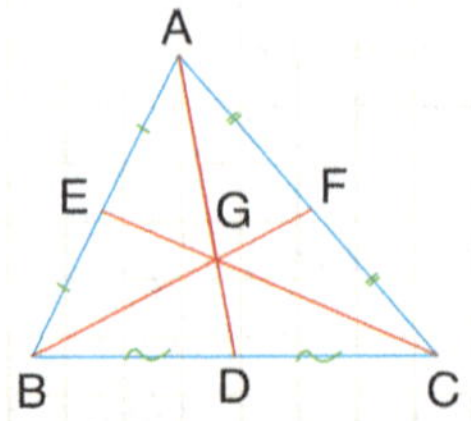

Example 1.13

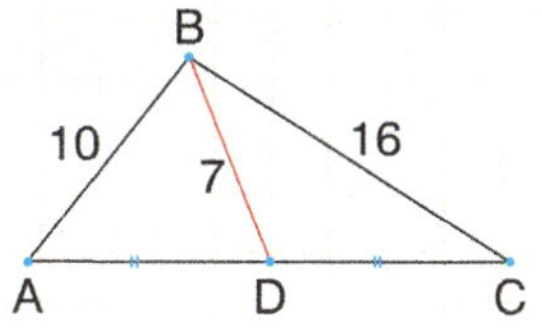

Area ABC?

Solution:

Draw DE || BC, meeting AB at E. Then ΔADE ~ ΔABC with ratio $\frac{1}{2}$.

Triangle BDE has sides 5, 7, 8.

$s = 10$, $[BDE] = \sqrt{10(5)(3)(2)} = 10\sqrt{3}$.

By similarity, $[ABC] = 4[BDE] = 4 \times 10\sqrt{3} = 40\sqrt{3}$.

Property 1.1

In the figure, let $|AE| = m$, $|EB| = n$, $|AF| = x$, $|FC| = y$, $|BD| = p$, $|DC| = q$.

Then, $\dfrac{A(DEF)}{A(ABC)} = \dfrac{mpy + nqx}{abc}$.

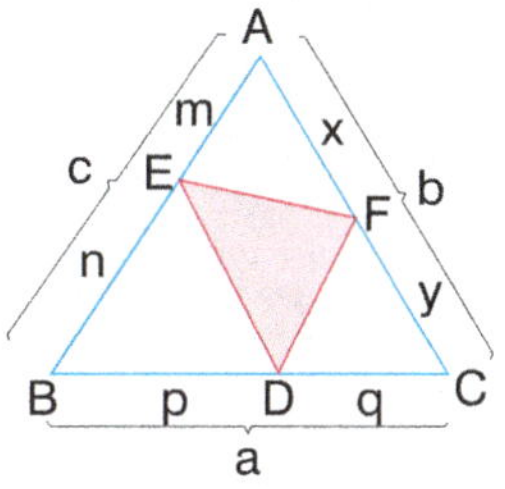

Example 1.14

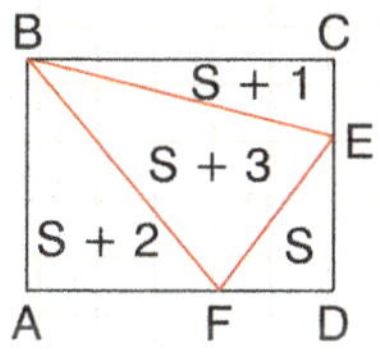

Rectangle ABCD is given. Points E and F lie on sides CD and AD, respectively. Segments BE and BF divide the rectangle into four regions:

- ΔABF has area s + 2.
- ΔBEC has area s + 1.
- ΔEFD has area s.
- ΔBEF has area s + 3.

Find the value of s.

Solution:

Let the rectangle have W and height H. Put AF = x on side AD and CE = y on side CD. Corner areas give:

$$[ABF] = \frac{1}{2}xH = s + 2 \Rightarrow xH = 2s + 4 \qquad (1)$$

$$[BEC] = \frac{1}{2}yW = s + 1 \Rightarrow yW = 2s + 2 \qquad (2)$$

Bottom-right triangle:

$$[EFD] = \frac{1}{2}(W - x)(H - y) = s \Rightarrow WH - xH - yW + xy = 2s. (3)$$

Total rectangle area: $WH = (s + 2) + (s + 1) + s + (s + 3) = 4s + 6$. (4)

From (3) with (1)(2)(4): $xy = 2s$. Now multiply (1) and (2):

$(xH)(yW) = (2s + 4)(2s + 2) = 4(s + 2)(s + 1)$.

But $(xH)(yW) = (xy)(WH) = (2s)(4s + 6)$.

Hence

$4(s + 2)(s + 1) = 2s(4s + 6) \Rightarrow$

$4s^2 + 12s + 8 = 8s^2 + 12s \Rightarrow s^2 = 2$. $s = \sqrt{2}$.

Example 1.15

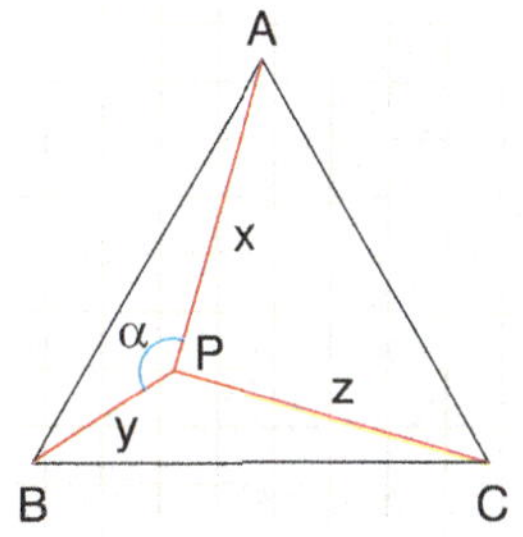

In a equilateral triangle ABC, a point P is taken inside the triangle such that

$$x^2 + y^2 = x^2.$$

Find the measure of $\angle APB = \alpha$.

Solution:

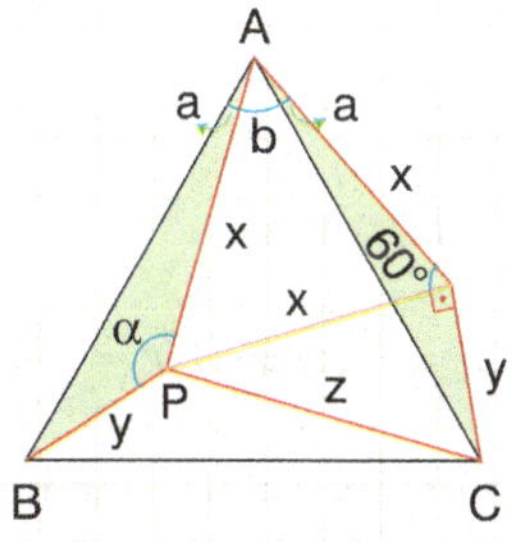

Construct triangle KAC so that ΔPAB ≅ ΔKAC.

Since A is a vertex of the equilateral triangle, $\angle BAC = 60°$.

Thus, $\angle PAB + \angle KAC = 60°$.

Draw segment PK. Then ΔAPK is equilateral, because $AP = PK = AK = x$.

In ΔPKC, we are given $x^2 + y^2 = z^2$, which implies $\angle PKC = 90°$.

Therefore,

$$\alpha = \angle APB = 60° + 90° = 150°.$$

Geometric View on Algebraic Problems

Algebraic identities often have hidden geometric meanings. For instance, the equation

$$x^2 + y^2 = a^2$$

can be seen as the Pythagorean relation in a right triangle with legs x, y and hypotenuse a.

A more surprising connection occurs if we look at

$$x^2 + xy + y^2 = a^2.$$

This comes directly from the Law of Cosines. If a triangle has sides x, y, a with the included angle between x and y equal to 120°, then

$$a^2 = x^2 + y^2 - 2xy\cos 120^\circ = x^2 + y^2 + xy.$$

So formula encodes the side relation of a 120° triangle.

System of Equations and Geometry

Now suppose positive numbers x, y, z satisfy:

$$\begin{cases} y^2 + yz + z^2 = b^2, \\ x^2 + xz + z^2 = c^2, \\ x^2 + xy + y^2 = a^2. \end{cases}$$

This corresponds to a triangle ABC with sides a, b, c, and an interior point P such that $\angle APB = \angle BPC = \angle CPA = 120^\circ$

The quantities xy, xz, yz represent scaled areas of the smaller triangles BPC, CPA, APB.

Expressing the Areas

Each smaller triangle has area:

$$[BPC] = \frac{1}{2}xy\sin 120^\circ = \frac{\sqrt{3}}{4}xy,$$

and similarly for the others. Adding them gives the total area of ΔABC:

$$\frac{\sqrt{3}}{4}(xy + xz + yz).$$

By Heron's formula, the area is also

$$\sqrt{p(p-a)(p-b)(p-c)}, \quad p = \frac{a+b+c}{2}.$$

Equating the two expressions, we find:

$$xy + xz + yz = \sqrt{\frac{(a+b+c)(a+b-c)(a+c-b)(b+c-a)}{3}}.$$

Example 1.16

Problem.

In a triangle with sides a = 7, b = 8, c = 9, an interior point P is chosen so that each angle at P is 120°. Compute xy + xz + yz.

Solution:

First compute the semiperimeter: p = (7 + 8 + 9) / 2 = 12.

Then Heron's formula gives area:

$$K = \sqrt{12\cdot(12-7)(12-8)(12-9)} = \sqrt{12\cdot 5\cdot 4\cdot 3}$$

$$= \sqrt{720} = 12\sqrt{5}.$$

So,

$$\frac{\sqrt{3}}{4}(xy + xz + yz) = 12\sqrt{5}.$$

Hence,

$$xy + xz + yz = \frac{48\sqrt{5}}{\sqrt{3}} = 16\sqrt{15}.$$

Example 1.17

Find

$$A = xy + 2yz + 3xz$$

given the system

$$3x^2 + 3xy + y^2 = 75, \ \ y^2 + 3z^2 = 48, \ \ x^2 + xz + z^2 = 9.$$

Solution:

Interpret with ABC having sides 3, 4, 5 and interior point P such that $\angle APB = 150^\circ$, $\angle BPC = 90^\circ$, $\angle CPA = 120^\circ$.

Areas of sub-triangles:

$$[APB] = \frac{xy}{4\sqrt{3}}, \ [BPC] = \frac{2yz}{4\sqrt{3}}, \ [APC] = \frac{3xz}{4\sqrt{3}}.$$

Thus,

$$[ABC] = 6 = \frac{1}{4\sqrt{3}}(xy + 2yz + 3xz).$$

So,

$$A = xy + 2yz + 3xz = 24\sqrt{3}.$$

Example 1.18

Positive Real Numbers

Let $x, y, z \in \mathbb{R}^+$ satisfy the following system of equations:

$$\sqrt{16-x^2} + \sqrt{25-y^2} + \sqrt{36-z^2} = 12$$

and

$$x + y + z = 9.$$

Find the value of:

$$xyz = ?$$

Solution:

$$\sqrt{4^2-x^2} + \sqrt{5^2-y^2} + \boxed{\sqrt{6^2-z^2}} = 12$$

$$x + y + z = 9$$

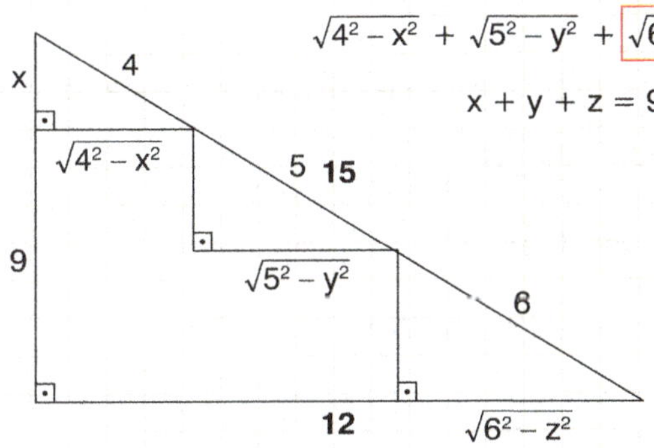

$$\sqrt{4^2-x^2} + \sqrt{5^2-y^2} + \sqrt{6^2-z^2} = 12, \quad x + y + z = 9$$

Geometrically, the three right triangles form a big right triangle with height 9 and base 12, so $\tan\theta = \frac{9}{12} = \frac{3}{4}$, giving $\sin\theta = \frac{3}{5}$, $\cos\theta = \frac{4}{5}$.

Then:

$$x = 4\sin\theta = \frac{12}{5},\ y = 5\sin\theta = 3,\ z = 6\sin\theta = \frac{18}{5}.$$

$$xyz = \frac{648}{25}$$

Example 1.19

In square **ABCD**, a line from vertex **A** intersects **CD** at **E** and **BC** at **F**.

Show that

$$\frac{1}{AE^2} + \frac{1}{AF^2} = \frac{1}{AB^2}$$

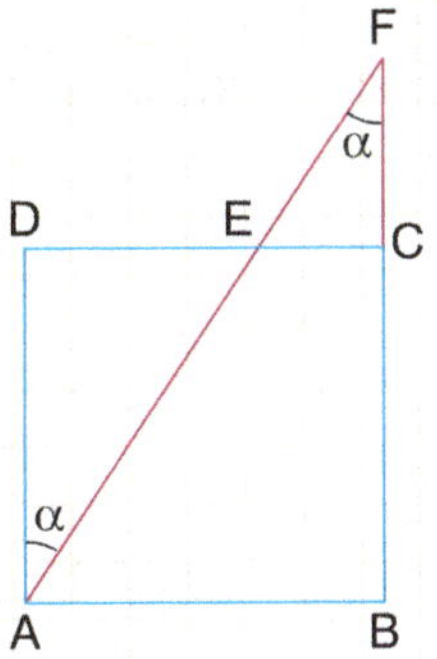

Solution:

In Δ**ADE**: $\cos\alpha = \frac{AD}{AE} = \frac{AB}{AE}$.

In Δ**ABF**: $\sin\alpha = \frac{AB}{AF}$.

Using $\cos^2\alpha + \sin^2\alpha = 1$:

$$\left(\frac{AB}{AE}\right)^2 + \left(\frac{AB}{AF}\right)^2 = 1.$$

$$AB^2\left(\frac{1}{AE^2} + \frac{1}{AF^2}\right) = 1.$$

Hence,

$$\frac{1}{AE^2} + \frac{1}{AF^2} = \frac{1}{AB^2}.$$

Example 1.20

- ABCD is a square
- E, F, G touch points
- AB = a, DE = b

Prove that: $\frac{a}{b} + \frac{b}{a} = 3$

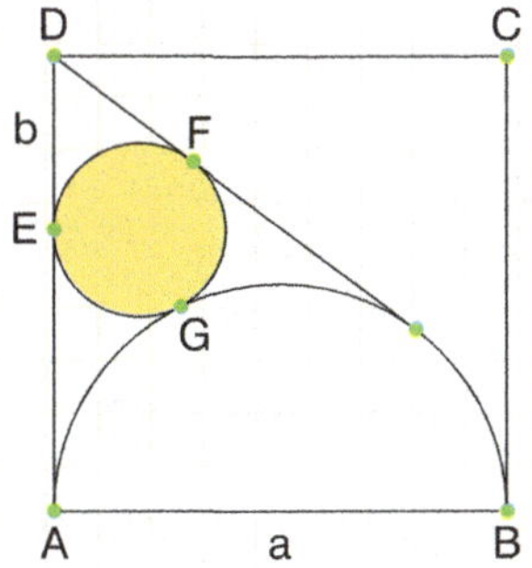

Solution:

1. ABCD is a square.

   The semicircle has radius $R = \frac{a}{2}$.

   The small circle touches AD at E with DE = b, and has radius r.

2. By similar right triangles at D:

$$\frac{b}{a} = \frac{r}{R} \Rightarrow r = \frac{b}{2}.$$

3. The circles are tangent, so the distance between centers equals R + r. By the same triangle similarity at D:

$$(a-b)^2 = 2ar.$$

4. Substitute $r = \frac{b}{2}$:

$$(a-b)^2 = ab \Rightarrow a^2 - 3ab + b^2 = 0.$$

Hence,

$$\frac{a}{b} + \frac{b}{a} = 3.$$

Example 1.21

In the figure, ABCDEFG is a regular heptagon.

Prove that $\frac{1}{a} = \frac{1}{b} + \frac{1}{c}$.

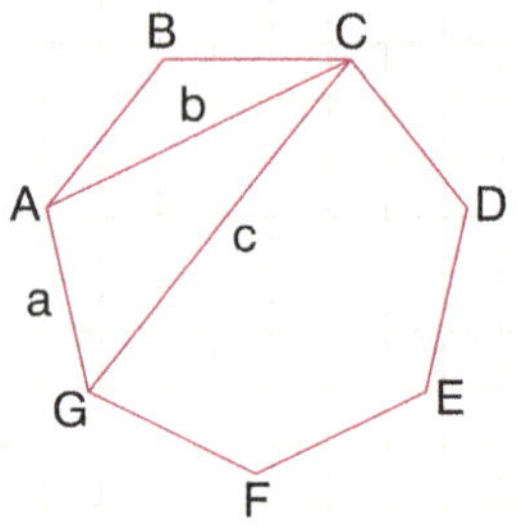

Solution:

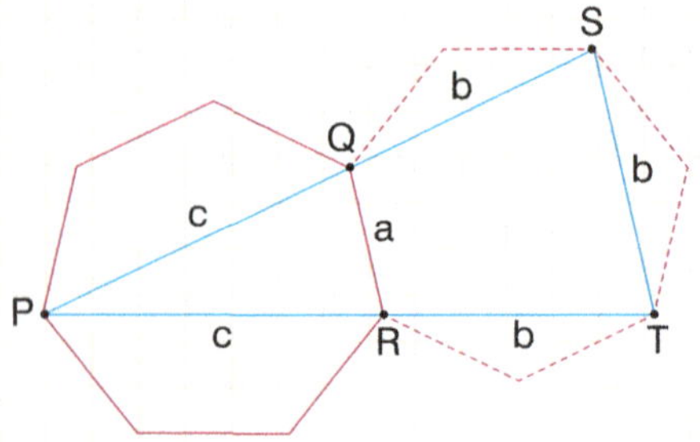

$\Delta PQR \sim \Delta PST$

$$\frac{c}{c+b} = \frac{a}{b} \Rightarrow cb = ac + ab \ (\div abc)$$

$$\frac{cb}{abc} = \frac{ac}{abc} + \frac{ab}{abc} \Rightarrow \frac{1}{a} = \frac{1}{b} + \frac{1}{c}.$$

Example 1.22

In right triangle ΔABC with $\angle A = 90°$, prove that

$$t_a = \frac{bc}{b+c}\sqrt{2},$$

where t_a is the bisector of $\angle A$.

Solution:

For any triangle,

$$t_a^2 = bc\left[1 - \left(\frac{a}{b+c}\right)^2\right].$$

When $\angle A = 90°$, we have $a^2 = b^2 + c^2$.

Substitube:

$$t_a^2 = bc\,\frac{(b+c)^2 - (b^2 + c^2)}{(b+c)^2} = bc\,\frac{2bc}{(b+c)^2} = \frac{2b^2c^2}{(b+c)^2}.$$

Hence,

$$t_a = \frac{bc}{b+c}\sqrt{2}.$$

Example 1.23

Prove that $\frac{a}{c} + (\frac{c}{a})^2 = 3$.

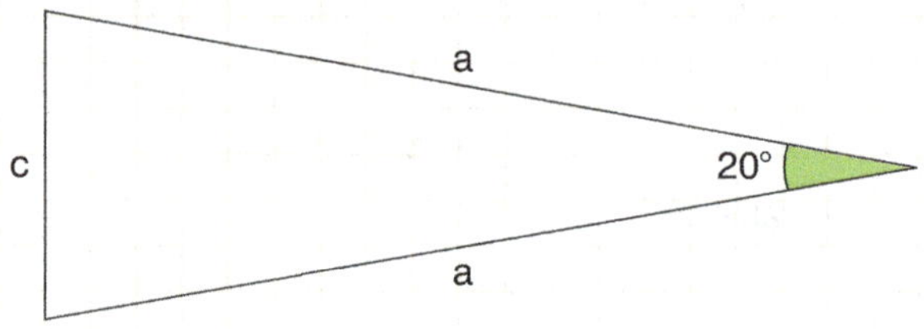

Solution:

We want $\frac{a}{c} + (\frac{c}{a})^2$ for isosceles triangle with sides a, a, c and angle 20° at the top.

Step 1. By Law of Cosines:

$$c^2 = 2a^2(1 - \cos 20°) = 4a^2 \sin^2 10°.$$

So

$$\frac{c}{a} = 2\sin 10°.$$

Step 2. Then

$$\frac{a}{c} = \frac{1}{2\sin 10°}, \quad (\frac{c}{a})^2 = 4\sin^2 10°.$$

Step 3. Sum:

$$\frac{1}{2\sin 10°} + 4\sin^2 10°.$$

Use triple-angle identity:

$$\sin 30° = \sin(3 \cdot 10°) = 3\sin 10° - 4\sin^3 10° = \frac{1}{2}.$$

So $4\sin^3 10° - 3\sin 10° = -\frac{1}{2}$.

Rearranging gives exactly the needed relation, yielding

$$\frac{1}{2\sin 10°} + 4\sin^2 10° = 3.$$

Practice 1

In triangle ABC, $\angle BAC = 2\angle ABC$, AB = 31, AC = 13. What is BC^2?

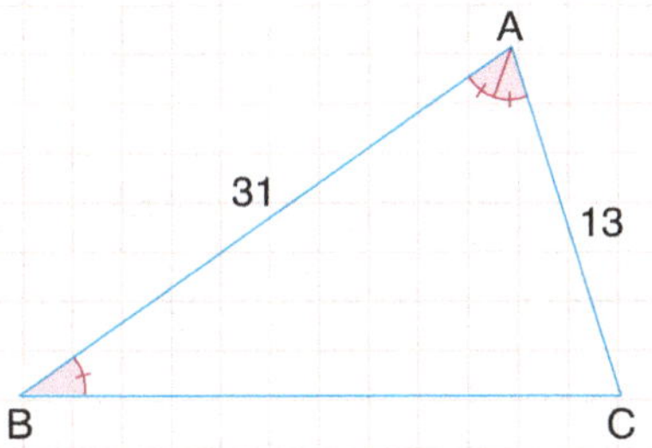

Solution:

From the rule of sines, we have

$$\frac{x}{\sin 2\alpha} = \frac{13}{\sin \alpha} = \frac{31}{\sin(180° - 3\alpha)} = \frac{31}{\sin 3\alpha}$$

$$4\cos^2 - 1 = \frac{\sin 3\alpha}{\sin \alpha} = \frac{31}{13}$$

$$2\cos\alpha = \frac{\sin 2\alpha}{\sin \alpha} = \frac{x}{13}$$

Squaring the second one and subtracting them, we get

$x^2 = 572.$

Practice 2

Triangle ABC is isosceles with AC = BC and $\angle ACB = 106°$ Point M is inside the triangle so that $\angle MAC = 7°$ and $\angle MCA = 23°$. Find the measure of $\angle CMB$ in degrees.

Solution:

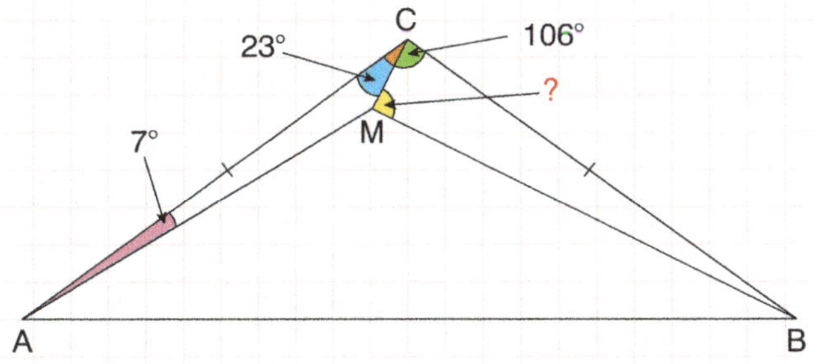

Practice 3

Let CMB = θ. Using Sine Rule, we have:

Solution:

$$\Delta ACM: \frac{CM}{\sin\angle MAC} = \frac{AC}{\sin\angle CMA} \Rightarrow \frac{CM}{\sin 7°} = \frac{AC}{\sin\angle CMA}$$

$$\Rightarrow CM = \frac{AC \sin 7°}{\sin 150°} = \frac{AC \sin 7°}{\sin 30°} = 2AC \sin 7°$$

$$\Delta BCM: \frac{CM}{\sin\angle MCB} = \frac{BC}{\sin\angle CMB} \Rightarrow \frac{CM}{\sin(97° - \theta)} = \frac{BC}{\sin\theta}$$

$$\Rightarrow CM = \frac{BC \cos(\theta - 7°)}{\sin\theta} = \frac{AC \cos(\theta - 7°)}{\sin\theta}$$

$$\Rightarrow 2\sin 7° = \frac{\cos(\theta - 7°)}{\sin\angle MCB}$$

$2\sin 7°\sin\theta = \cos(\theta - 7°)$

$2\sin 7°\sin\theta = \cos 7°\cos\theta + \sin 7°\sin\theta$

$\cos 7°\cos\theta - \sin 7°\sin\theta = 0$

$\cos\theta + 7° = 0$

$\Rightarrow \theta + 7° = 90°$

$\Rightarrow \angle CMB = \theta = 90° - 7° = 83°$

Practice 4

Let ABCDE be a convex pentagon such that AB = AE = CD = 1, $\angle ABC = \angle DEA = 90°$ and BC + DE = 1. Compute the area of the pentagon.

(JBMO 1998)

Solution:

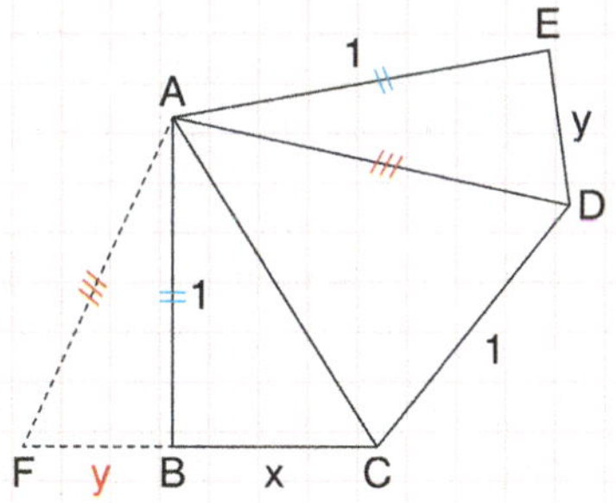

Let's draw $\widehat{AFB} \cong \widehat{ADE}$.

AF = AD, BF = ED = y.

Therefore

CF = x + y = 1 = CD

and AFCD is a kite.

Hence,

$\text{Area}(ABCDE) = \text{Area}(AFCD) = 2 \cdot \text{Area}(\widehat{AFC}) = 2 \cdot (\frac{1 \cdot 1}{2})$

$= 1.$

Practice 5

ABCD is a parallelogram. Point E on the side $\overline{AB}$ and point F on the side $\overline{BC}$ such that Area(ADE) = Area(DCF) = 10 and Area(EBF) = 9. Find Area(DEF).

(L. Gökçe, Antalya City M.O 2012)

Solution:

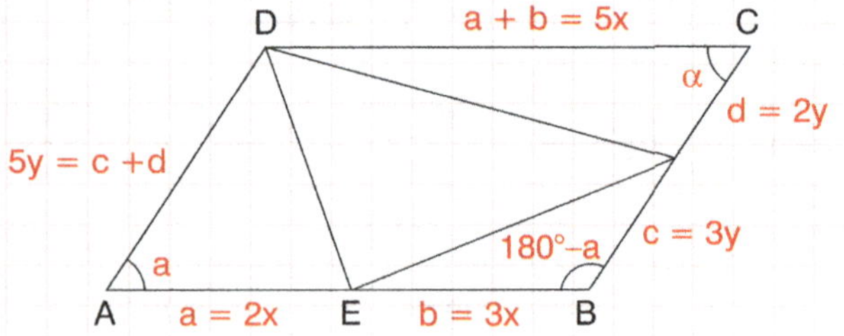

By sine area relation, $\frac{1}{2} \cdot a(c + d) \cdot \sin\alpha =$

$= \frac{1}{2} d(a + b) \cdot \sin\alpha = 10$ and $\frac{1}{2} bc \cdot \sin(180° - \alpha)$ 9.

Hence, ac + a$\not{d}$ = a$\not{d}$ + bd and 9 · (ac+ ad) = 10 bc.

Then, $\frac{a}{b} = \frac{d}{c} = t$ and $9 \cdot \left(\frac{ac}{bc} + \frac{ad}{bc}\right) = 10$

$\Rightarrow 9t^2 + 9t - 10 = 0 \Rightarrow (3t + 5)(3t - 2) = 0 \Rightarrow t = \frac{2}{3}$.

3t +5
3t –2

Let a = 2x, b = 3x, cos3y, d= 2y.

Area(ADE) $= \frac{1}{2} \cdot 2x \cdot 5y \cdot \sin\alpha = 10 \Rightarrow xy\sin\alpha = 2$. Thus,

Area(ABCD) $= 5x \cdot 5y \cdot \sin\alpha = 25 \times 2 = 50$ and

Area(DEF) = 50 – 10 – 10 – 9= 21.

My Notes:

Practice 6

Positive x, y, z satisfy a system:

$$\begin{cases} x^2 + xy + \frac{y^2}{3} = 25 \\ \frac{y^2}{3} + z^2 = 9 \\ z^2 + zx + x^2 = 16 \end{cases}$$

Find the value of expression xy + 2yz + 3zx.

Solution:

Given x, y, z > 0 and we remember statement of the Law of cosines theorem. Let's draw a triangle ABC that AB = 3, BC = 5, CA = 4. Let's take a point P interior of the triangular region such that PA = z, PB = $\frac{y}{\sqrt{3}}$, PC = x and ∠APB = 90°,

∠APC = 120°, ∠PBC = 150°. Now,

Area (ABC) $= \frac{3 \square 4}{2} = 6$ and by area formula with sine,

Area (ABC) $= \frac{1}{2}\square\left[\frac{\sqrt{3}xy}{2} + \frac{zy}{\sqrt{3}} + \frac{\sqrt{3}xz}{2}\right] = 6$.

Hence, this yeilds $xy + 2yz + 3zx = 24\sqrt{3}$.

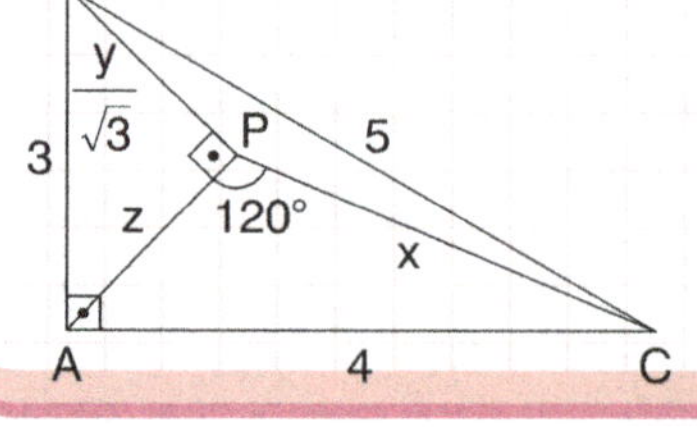

My Notes:

MathTopia Press

Practice 7

In the diagram below, ABCD is a rectangle with side lengths AB = 3 and BC = 11, and AECF is a rectangle with side lengths AF = 7 and FC = 9, as shown. The area of the shaded region common to the interiors of both rectangles is $\frac{m}{n}$, where m and n are relatively prime positive integers. Find m + n.

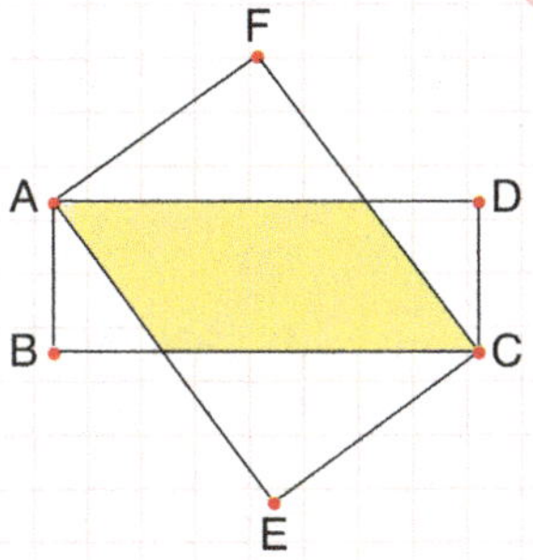

Work Space

Practice 8

In triangle ABC, $m(\angle ABC) = \alpha$, $m(\angle BAC) = 90° + \alpha$.

If $|AB| = 28$ cm and $|AC| = 60$ cm,

find the length of $|BC|$.

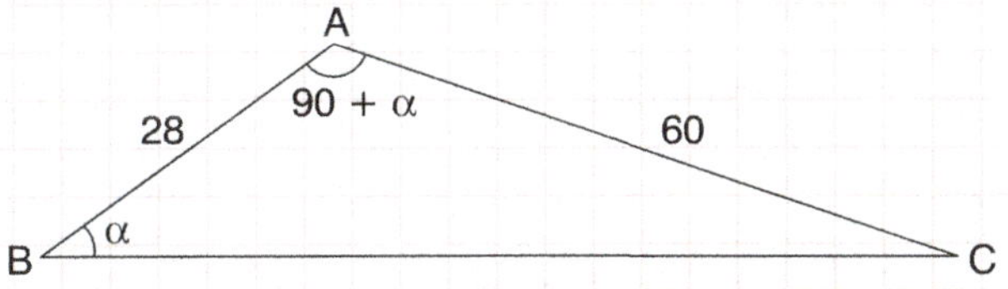

Work Space

Practice 9

In triangle **ABC**, let **D** be the midpoint of **BC**.

If **|AE|** = **|AF|** and points **F**, **D**, **E** are collinear, show that

$$|AF| = \frac{|AB| + |AC|}{2}.$$

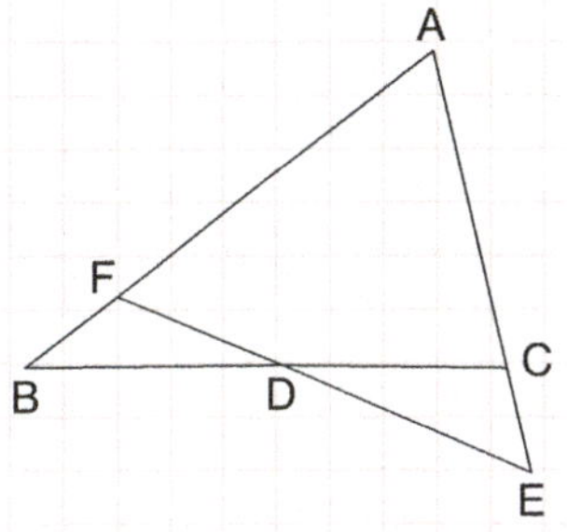

Work Space

Practice 10

Let a and b be positive real numbers such that a + b = 15.

Find the minimum possible value of

$$\sqrt{a^2 + 9} + \sqrt{b^2 + 25}.$$

Work Space

My Notes:

My Notes:

1. In a convex quanrilateral, the diagonals are perpendicular. If the sum of the lengths of the diagonals is 12, what is the maximum possible area of the quadrilateral?

2. A quadrilateral has side lengths 1, 4, 7, and 8. What is the maximum possible area of the quadrilateral?

A) $7\sqrt{2}$ B) $10\sqrt{3}$ C) 18 D) $12\sqrt{3}$ E) $9\sqrt{5}$

3. Suppose x, y and z are real numbers such that

$x^2 + y^2 = 9$,

$y^2 + z^2 = 16$, and

$y^2 = x \cdot z$.

What is $x + y + z$?

4. Take a square ABCD of side length 1, and let P be the midpoint of AB. Fold the square so that point D touches P, and let the intersection of the bottom edge DC with the right edge be Q. What is BQ?

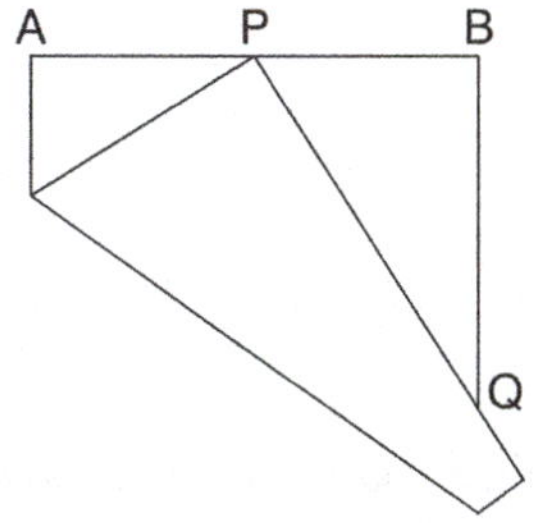

5.

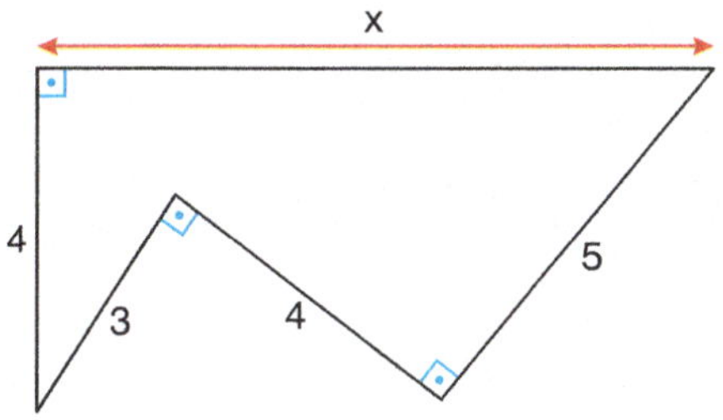

6.

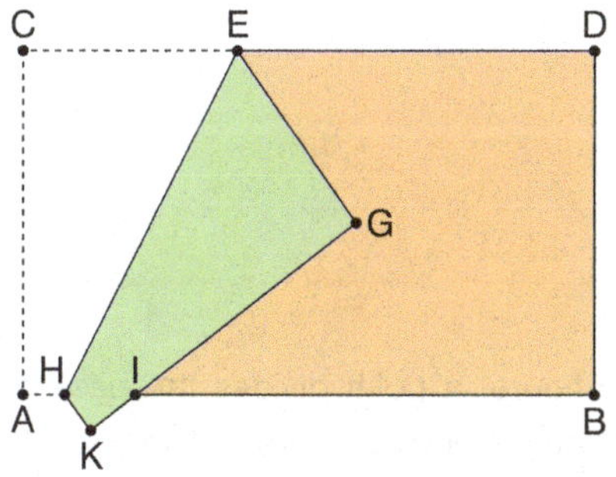

AH = 3

AC = 24

CE = 15

K, I, G, D collinear

Perimeter(EGIBD) = ?

7. Two altitudes of a triangle have lengths 8 and 12. Which of the following cannot be the length of the third altitude?

A) 4 B) 7 C) 8 D) 12 E) 23

8. A piece of cardboard measuring 40 x 80 is folded along the diagonal as show.

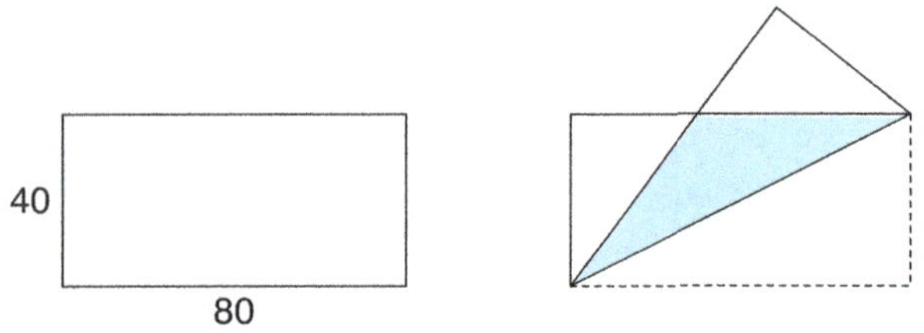

What is the area of the overlap?

9.

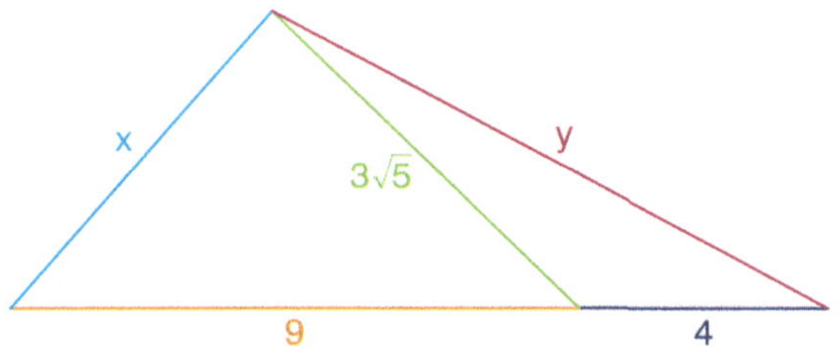

In the figure, a point divides the base of a triangle into segments of lengths 9 and 4. The line from the top vertex to this point has length $3\sqrt{5}$. If the other two sides of triangle are x and y, what is the maximum possible value of x + y?

10. A sheet of paper 12 inches by 18 inches is folded so that two opposite corners touch, as shown in the figures below. What is the area, in square inches, of the shaded triangle formed where the paper overlaps itself?

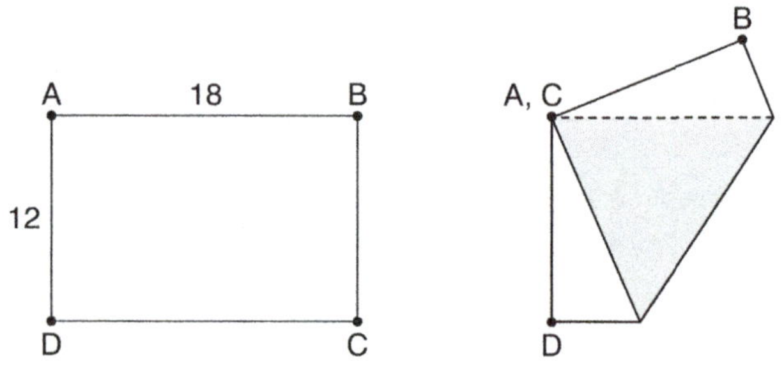

11. Find the maximum value of the expression $A = x + z$ if x, y, z, w satisfy the following system $\begin{cases} x^2 + y^2 = 4 \\ z^2 + t^2 = 9 \\ xt + yz = 6. \end{cases}$

MathTopia Press

12. **2021 AMC 10B, #21)**

A square piece of paper has side length 1 and vertices A, B, C, and D in that order. As shown in the figure, the paper is folded so that vertex C meets edge $\overline{AD}$ at point C', and edge $\overline{BC}$ intersects edge $\overline{AB}$ at point E. Suppose that $C'D = \frac{1}{3}$. What is the perimeter of triangle $\triangle AEC'$?

A) 2 B) $1+\frac{2}{3}\sqrt{3}$ C) $\frac{13}{3}$ D) $1+\frac{3}{4}\sqrt{3}$ E) $\frac{7}{3}$

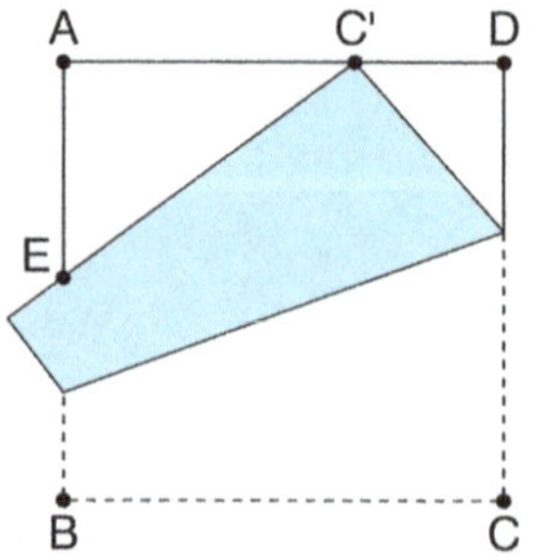

1. In a convex quadrilateral ABCD, the diagonals intersect at O. If the areas of $\triangle AOB$ and $\triangle COD$ are 4 and 9, respectively, what is the least possible area of quadrilateral ABCD?

A) 20 B) 22 C) 24 D) 25 E) 27

2. A quadrangular shape is given. The sides are divided by midpoints. What is the area of the fourth region?

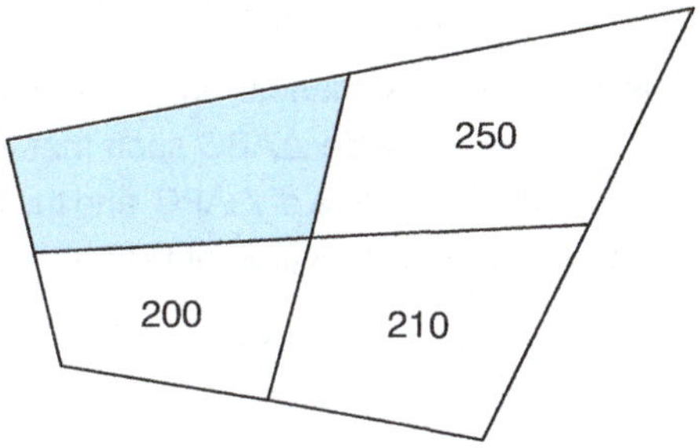

3. A regular heptagon is divided as shown, where each segment is labeled a, b, c, or 1. Determine the value of

$a^2 + b^2 + c^2$.

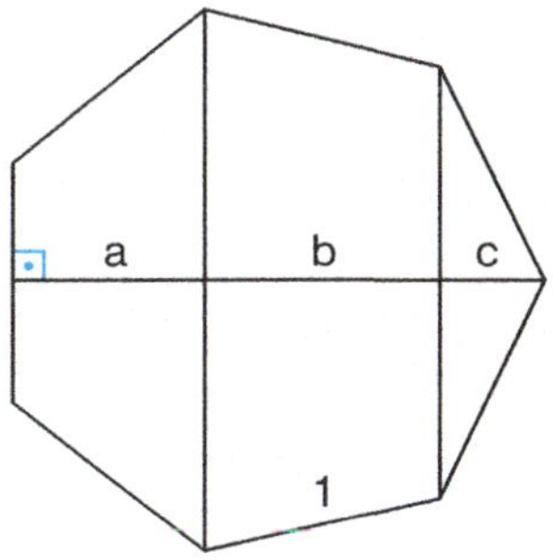

4. Let $L_1 : 3x + 4y = 10$, $L_2 : x - 2y = 0$, and $L_3 : x = 0$. These three lines form a triangle. What is the distance from the incenter of this triangle to line L_1?

A) $\frac{\sqrt{5}-1}{2}$ B) $\frac{5-\sqrt{5}}{4}$ C) $\frac{\sqrt{5}}{4}$ D) $\frac{3-\sqrt{5}}{2}$ E) $\frac{2}{3}$

5. **(CEMC, 2021, Problem of the Week)**

Quadrilateral BDFH is constructed so that each vertex is on a different side of square ACEG. Vertex B is on side AC so that AB = 4 cm and BC = 6 cm. Vertex F is on EG so that EF = 3 cm and FG = 7 cm. Vertex H is on GA so that GH = 4 cm and HA = 6 cm. The area of quadrilateral BDFH is 47 cm^2.

Find CD?

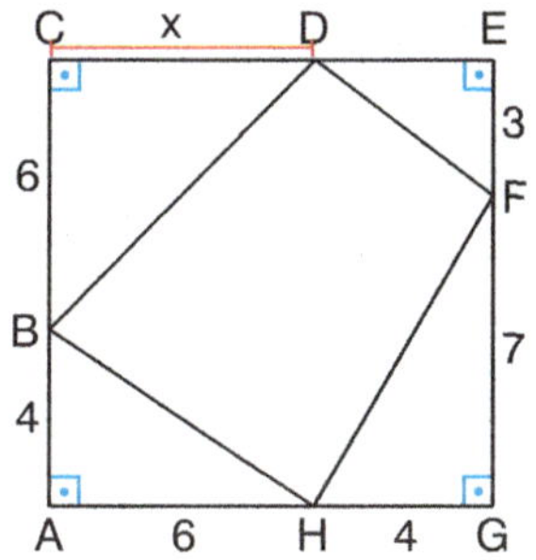

6. An isosceles trapezoid has perimeter 28. Its longest side has length 13. What is the maximum possible area of the trapezoid?

7.

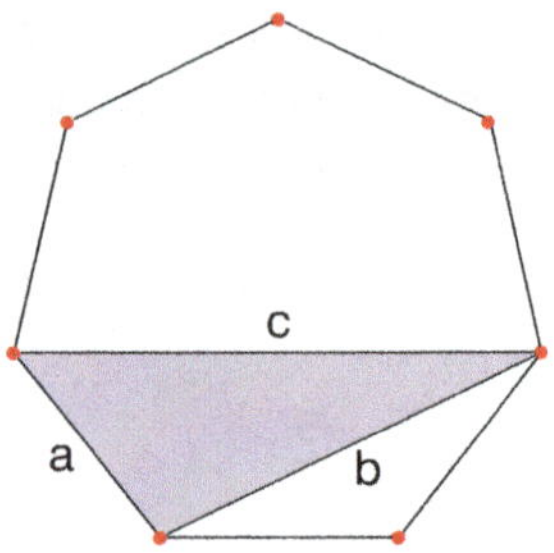

$$\frac{a^2}{b^2} + \frac{b^2}{c^2} + \frac{c^2}{a^2} = ?$$

8.

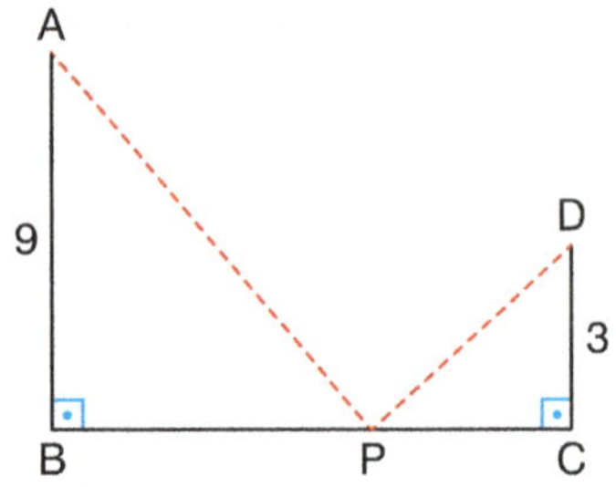

[AB] ⊥ BC

[DC] ⊥ BC

|AB| = 9 cm

|BC| = 8 cm

|DC| = 3 cm

Since, for a point P taken on the line BC, ||AP| – |DP|| of the statement

What is the largest value in cm?

9. Suppose real numbers x, y satisfy

$$\sqrt{x^2 + 1} + \sqrt{y^2 + 4} = 7.$$

What is the maximum possible value of x + y?

10. In △ABC with sides 5, 12, 13, an interior point P satisfies

$$\angle APB = 120°, \angle BPC = 150°, \angle CPA = 90°.$$

Let $AP = x$, $BP = \frac{y}{\sqrt{3}}$, $CP = z$ (with $x, y, z > 0$).

Evaluate

$$S = xy + \frac{1}{\sqrt{3}}yz + 2xz.$$

11. ABC is an isosceles triangle with AB = AC = 5 and BC = 8. Point P is inside △ABC such that the product of the area of △APB, the area of △APC, and the area of △BPC is maximum. What is the value of PB?

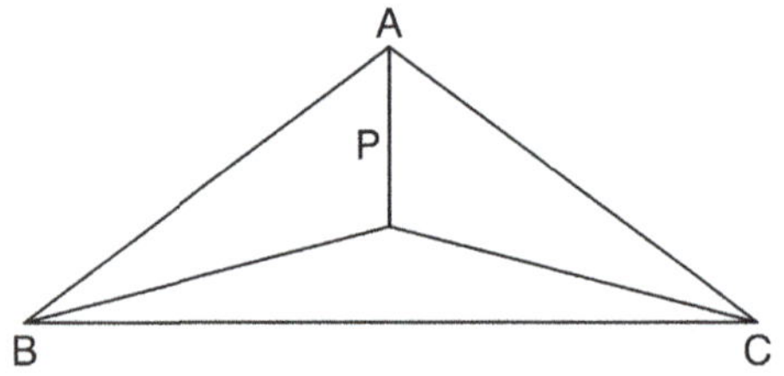

A) 3 B) $\sqrt{10}$ C) $2\sqrt{3}$ D) $\sqrt{17}$ E) $2\sqrt{5}$

12. Let

$A = xy + 2yz + 2xz,$

where positive numbers x, y, z satisfy

$$\begin{cases} 3x^2 + 3xy + y^2 = 75, \\ y^2 + 3z^2 = 48, \\ x^2 + xz + z^2 = 9. \end{cases}$$

Find A.

GeoTopia

AMC 10 & 12 PRACTICES

- Polygons, Triangles & Circles
- Inequalities & Maximum Area
- Out-of-the-Box Connections

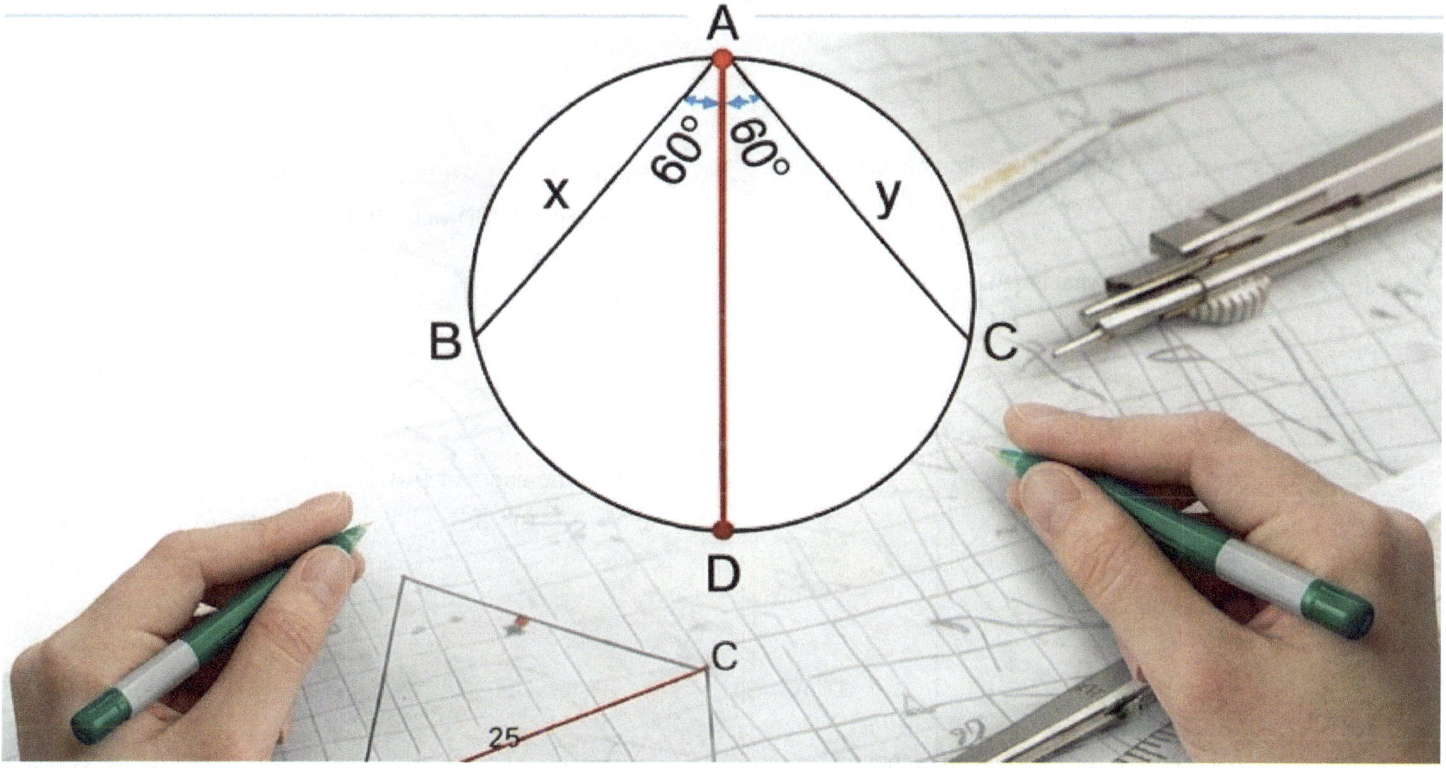

Tailored for Intro AIME-Level Students

AMC 10/12 • AIME • Euclid

1.

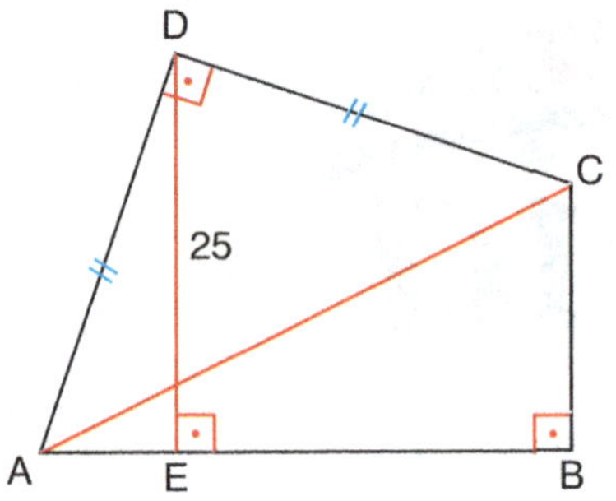

In quadrilateral ABCD, it is given that $\angle DAB = \angle ABC = \angle ADC = 90°$. Point E lies on AB such that $AE \perp DE$.

Suppose AD = DC, and the length of DE is 25. What is the length of diagonal AC?

A) $25\sqrt{2}$ B) 40 C) 45 D) $50\sqrt{2}$ E) 75

2. In the circle shown, point A lies on the circumference, and chords AB, AC, and AD are drawn such that $\angle BAC = 120°$, and $\angle BAD = \angle CAD = 60°$.

If $|AB| = x$ and $|AC| = y$, what is $|AD|$ in terms of x and y?

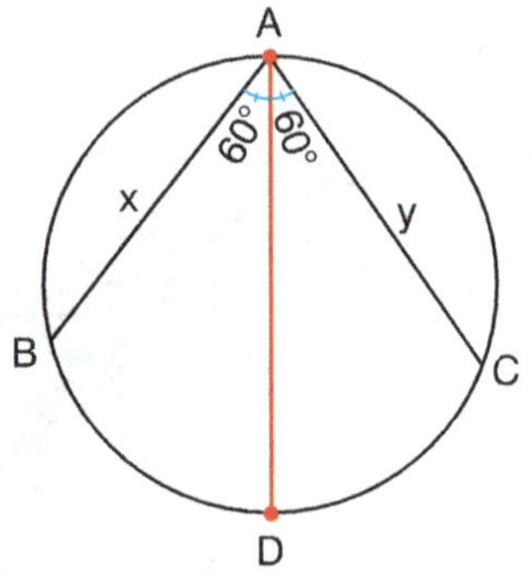

A) $2(x + y)$ B) $x + y$ C) $x - y$
D) $y - z$ E) None of these

3. In trapezoid ABCD with AB | | CD, points K and L are chosen on side [AB] such that K lies between A and L, and the following conditions are satisfied:

$$|2x| = 2|CD| = 3|BL| = 4|AK|.$$

It is given that the area of trapezoid ABCD is 40, and the area of triangle BCD is 12.

Find the area of triangle KLC.

A) 10 B) 12 C) 14 D) 16 E) 18

4. **(TÜBİTAK 2021, High School Math Olympiad)**

In trapezoid ABCD with AB| |CD:

- $|CD| = 6$,
- $|AC| = 3\sqrt{2} + \sqrt{6}$,
- $|BC| = 2\sqrt{3} + 2$,

and it is given that

$$\angle DAC = \angle DCB.$$

What is the ength of |AB|?

A) 3 B) $\sqrt{5}$ C) $\sqrt{3}$ D) $\sqrt{6} - 1$ E) 2

5. In rectangle ABCD, point E is located inside the rectangle such that

$$ED = 2\sqrt{2},\ EB = 2\sqrt{3},\ EC = \sqrt{3},$$

and the side lengths satisfy

$$BC = \sqrt{3} \cdot DC.$$

What is the measure of $\angle BEC$?

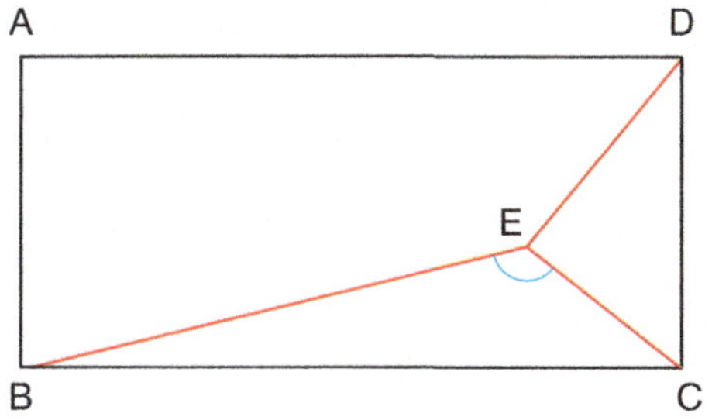

A) 150 B) 140 C) 135 D) 125 E) 120

6. (2025 AMC 8 #24)

In trapezoid ABCD, angles B and C measure 60° and AB = DC. The side lengths are all positive integers, and the perimeter of ABCD is 30 units. How many non-congruent trapezoids satisfy all of these conditions?

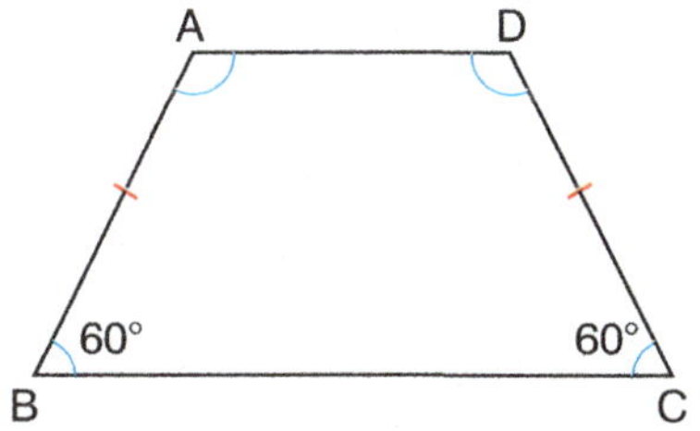

A) 0 B) 1 C) 2 D) $\sqrt{3}$ E) 4

7. (TÜBİTAK 2019 Middle School Math Olympiad)

In right triangle ABC, where $\angle ABC = 90°$, point D is chosen on side [AC]. It is given that:

$$\frac{|AB|}{|BC|} = \sqrt{3},\ \frac{|BD|}{|AC|} = \frac{\sqrt{3}}{4}$$

What is the measure of $\angle BDC$?

A) 30° B) 45° C) 60° D) 75° E) 90°

8. (TÜBİTAK 2024 Middle School Math Olympiad)

In an acute triangle ABC, a point D is chosen inside the triangle such that

$$\angle DBC = \angle BAC = 45°,$$

and it is given that

$$|BD| = 5,\ |BC| = 5\sqrt{2}.$$

What is the length of $|AD|$?

A) 3 B) 4 C) $5\sqrt{2}$ D) 10 E) None of these

9. In convex pentagon ABCDE,

$\angle ABC = \angle DEA = 90^\circ$, $AB = CD = EA = 1$.

If $BC + ED = 1$, then what is the area of pentagon ABCDE?

A) $\frac{1}{4}$ B) $\frac{1}{3}$ C) $\frac{1}{2}$ D) 1 E) $\frac{3}{2}$

10. **(TÜBİTAK 2021 Middle School Math Olympiad)**

In ΔABC, points D and E are chosen on side [BC] such that E lies between C and D. Lines AD and AE are extended to meet the circumcircle of ΔABC afain at points K and L, respectively.

It is given that:

$|BD| = 2$, $|DK| = 3$, $|DE| = 4$, $|EC| = |EA|$, and $|EL| = |DA|$.

What is the length of $|AL|$?

A) 7 B) 8 C) 9 D) 10 E) 11

11. A semicircle has diameter $AB = 2r$. Points A, P, E, B lie on the semicircle in that order. The chord lengths satisfy

$PA = 2$, $PE = 2$, $BE = 6$.

What is the value of $r^3 - 11r$?

A) 0 B) 3 C) 4 D) 6 E) 8

MathTopia Press

12. Let P be a point inside equilateral triangle ABC such that

$m\angle APB = 150^\circ$, $AP = 2\sqrt{3}$, $BP = 2$.

Find PC.

A) 2 B) 3 C) 4 D) $2\sqrt{7}$ E) $3\sqrt{3}$

1. In trapezoid ABCD with AB || CD, AB = 6, BC = 3. Point E lies on line CD. If ∠EBC = ∠EBA and BE = 5, what is the length of AE?

A) 0 B) $\sqrt{7}$ C) $\sqrt{11}$ D) 4 E) 5

2. Square ABCD has a point E on side BC such that ∠EAB = 15°. From C, a perpendicular is dropped to line AE, meeting at H. If CH = 2, what is the area of the square?

A) 4 B) 6 C) 8 D) 12 E) 16

3. **(Antalya Matematik Olimpiyatı 2019, #1)**

In ΔABC, points D on AC, E on BC, and F on AE are chosen. The areas of three small triangles are give:

[AFD] = 1, [ABF] = 2, [BEF] = 3.

What is the area of quadrilateral CDFE?

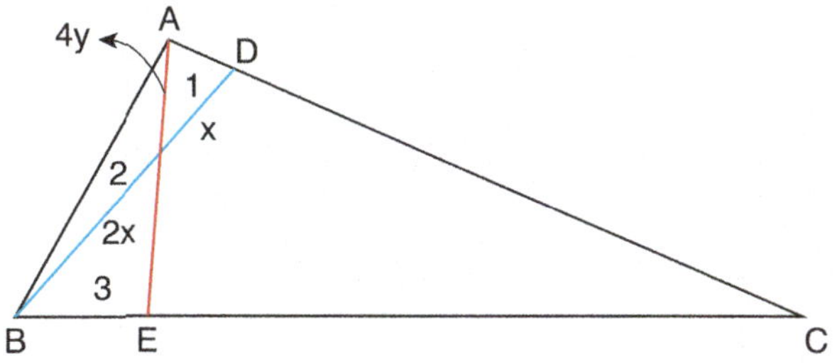

A) 24 B) 26 C) 21 D) 22 E) 25

4. **(TÜBİTAK 2019 Middle School Math Olympiad, #5)**

In ΔABC, points D lies on side BC. A line parallel to AC passes through D, intersecting AB at F and AC at E. If

[DFB] = 1, [DEF] = 2, [DEC] = 3,

what is the area of [AFE]?

A) 1 B) $\frac{3}{2}$ C) 2 D) $\frac{5}{2}$ E) 3

5. In right triangle ABC with $\angle B = 90°$, let BD be the median to hypotenuse AC. Extend BD to meet line d, which passes trough A and is perpendicular to AC, at point E.

If $\angle AEB = 18°$ and $AB = 12$, what is the length of DE?

A) 12 B) 18 C) 20 D) 24 E) 30

6. A semicircle has diameter $AB = 2r$. Points A, P, E, B lie on the semicircle in that order. The chords satisfy

$PA = 2$, $PE = 2$, $BE = 2$.

What is the value of r?

A) 1 B) $\frac{3}{2}$ C) 2 D) $\sqrt{5}$ E) 3

7. (TÜBİTAK 2019 Middle School Math Olympiad, #13)

In convex quadrilateral ABCD:

$\angle ABC = 45°$, $\angle ADC = 90°$, $\angle BAC = \angle CAD$,

with $BC = 2$, $CA = \sqrt{3}$, find AD.

A) $\frac{\sqrt{3}}{2}$ B) $\sqrt{2}$ C) $\frac{\sqrt{3}}{2}$ D) 1 E) $\frac{1}{2}$

8. (TÜBİTAK 2021 Middle School Math Olympiad, #21)

Trapezoid ABCD with AB || CD is given. Point F lies on AB. The perimeters of triangle ΔAFD, ΔFDC, ΔFCB are all equal. If $AB = 12$, find the sum of all possible values of CD.

A) 4 B) 6 C) 8 D) 10 E) 12

9. Square ABCD has side length s. A point P is located inside the square such that

PA = 2, PB = 5, PD = $\sqrt{17}$.

The area of the square can be expressed in the form m + n $\sqrt{p}$, where m, n, p are positive integers and p is square-free.

Find m + n + p.

A) 49 B) 53 C) 55 D) 57 E) 59

10. In triangle ADE, points B on AE and C on DE are chsen so that quadrilateral ABCD is cyclic. Let BD and AC intersect at F. If ∠EAC = 21° and ∠AED = 33°, what is the measure of ∠AFD?

A) 63° B) 69° C) 72° D) 75° E) 78°

11. **(TÜBİTAK 2021 Middle School Math Olympiad, #29)**

Trapezoid ABCD has AB || CD. Two circles ω_1, ω_1 are tangent externally inside the trapezoid. Circle ω_1 is tanget to sides AB, CD, DA. Circle ω_2 is tangent to sides AB, BC, CD.

If the area of trapezoid ABCD s 12 and |AD| + |BC| = 8, find |AB| + |CD|.

A) 8 B) 10 C) 12 D) 14 E) 16

12. Let ABC be a triangle with AB= 28. Points D and E lie on side BC such that BD = DE = EC.

Point K lies on AB with BK = 16 and KA = 12. Point L lies on AC with CL = 6 and AL = x.

It is given that

$$\angle BKE = \angle CLD.$$

Find the value of x.

A) 8 B) 9 C) 10 D) 11 E) 12

1. Let C ∈ [AB]. On the semicircle with diameter [AB], point D and E are taken such that $\angle ACD = \angle BCE = 30°$.

If $|AC| = 27$, $|CB| = 3$, then what is $|CD| - |CE|$?

A) 15 B) 18 C) 21 D) $10\sqrt{3}$ E) $12\sqrt{3}$

2. In right triangle ΔACE, the side engths are AC = 12, CE = 16, and EA = 20. Points B, D, and F lie on sides AC, CE, and EA, respectively, such that AB = 3, CD = 4, and EF = 5.

What is the ratio of the area of ΔDBF to the area of ΔACE?

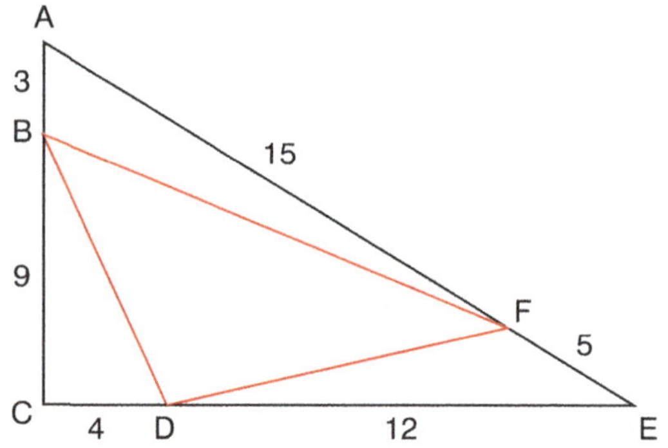

A) $\frac{3}{16}$ B) $\frac{5}{16}$ C) $\frac{7}{16}$ D) $\frac{9}{16}$ E) $\frac{1}{2}$

3. A semicircle has diameter AB = 2r. Points A, P, E, B lie on the semicircle in that order. The chord lengths satify PA = 2, PE = 2, BE = 2.

What is r^2?

A) 3 B) 4 C) 5 D) 6 E) 7

4. Let AB be the diameter of a semicircle and let C be apoint on the semicircle. A chord from C to B has length CB = 7 cm. From A toward C along the circle, two consecutive equal chords each have length 2 cm. Find the radius R of the semicircle.

A) 3 B) 4 C) 5 D) 6 E) 7

5. **(TÜBİTAK 2021 Middle School Math Olympiad, #9)**

In right triangle ABC with $\angle BAC = 90°$, the side lengths are $|AB| = 1$ and $|AC| = 2$.

A point D is chosen on side [BC], and a point. E is chosen on segment [CD] such that

$|AD| = \frac{2}{\sqrt{5}}$, and Area ($\Delta DAE$) = Area ($\Delta ACE$).

Find the length $|AE|$.

A) $\frac{\sqrt{10}}{7}$ B) $\frac{\sqrt{2}}{2}$ C) $\frac{3\sqrt{10}}{2}$ D) $\frac{2\sqrt{10}}{5}$ E) $2\sqrt{10}$

6. **(2005 AIME II, Problem 8)**

Circles ω_1 and ω_2 are externally tangent, and both are internally tangent to circle Ω. The radii of ω_1 and ω_2 are 4 and 10, respectively, and the centers of all three circles are collinear. A chord of Ω is also a common external tangent to ω_1 and ω_2.

If the length of the chord can be written in the form

$$\frac{m\sqrt{n}}{p},$$

where m, n, p, are positive integers, gcd (m, p) = 1, and n is squarefree, compute m + n + p.

7. In ΔABC, points D on side BC and E on side AC are chosen. Segments AD and BE intersect at point P.

The triangle is divided into four smaller regions. Their areas are given as follows:

[APE] = 4, [BPD] = 9, [APB] = 16.

Find the area of the remaining region.

A) 22 B) 24 C) 30 D) 36 E) 40

8. **(TÜBİTAK 2019 Middle School Math Olympiad, #21)**

In ΔABC, AB = 11, BC = 17. Point D is chosen on BC so that DC = 3. Let E be the midpoint of AC. If $\angle BDE = 56°$, find $\angle ABC$.

A) 56° B) 60° C) 64° D) 68 E) 72°

9. If ΔABC, point D lies on side BC, and point E lies on segment AD. It is given that

$DE = 1$, $AE = 2$, $BD = CD = \sqrt{3}$.

What is $\angle BAC + \angle BEC$?

A) 90° B) 120° C) 135° D) 150° E) 180°

10. **(2023 AIME I, Problem 5)**

Let P be a poit on the circle circumscribing square ABCD such that

$PA \cdot PC = 56$ and $PB \cdot PD = 90$.

Find the area of ABCD.

11. **(TÜBİTAK 2020 Middle School Math Olympiad - First Round)**

In ΔABC, $|AC| = |AB| = 25$ and $|BC| = 40$. Let D be the midpoint of [BC], and et E be the foot of the perpendicular from B to AC.

Find the diameter of the circle that passes through D and is tangent to line AC at E.

A) 30 B) 32 C) $\frac{100}{3}$ D) 34 E) 36

12. Triangle ABC has side lengths $AB = 13$, $BC = 14$, and $AC = 15$. Let J_A denote the center of the A-excircle of ΔABC (the circle tangent to line BC and the extensions of AB and AC). What is the length AJ_A?

A) $6\sqrt{15}$ B) $\sqrt{541}$ C) $9\sqrt{7}$ D) 24 E) $3\sqrt{65}$

SOLUTIONS

ANSWERS

1. 32°. Angle between bisectors:

$$\frac{16°}{2} + \frac{48°}{2} = 32°$$

2. 165. Isosceles cases give x = 40, 55, 70.

3. 80°. Let ∠B = 3p, ∠C = 3q.

$30 + 3p + 3q = 180 \Rightarrow p + q = 50$

$\angle BDC = 180 - 2(p + q) = 80°$

4. 117°. LAB = BE and ∠B = 60° imply ΔABE equilateral.

$$\angle EAC = 114° - 60° = 54°$$

$$\angle EDC = 180 - \frac{180 - 54}{2} = 117°$$

5. 2. Given ∠B + ∠C < 150:

$$6x + 15 < 150 \Rightarrow x \le 22$$

Since AB > BC ⇒ ∠C > ∠A:

$$x + 25 > 165 - 6x \Rightarrow 7x > 140 \Rightarrow x \ge 21$$

Also $\angle B > 0 \Rightarrow 5x - 10 > 0 \Rightarrow x \ge 3$ (already satisfield).

So $x \in \{21, 22\} \Rightarrow 2$.

6. 11. $\frac{1}{2} \cdot 6(y - 7) = 12 \Rightarrow y = 11$

7. 13. Using coordinates from perpendicular sides,

$$DC = \sqrt{12^2 + 5^2} = 13$$

8. $\sqrt{2}$. Isosceles right triangle with BH = 5, HC = 3 gives

$$AD = \sqrt{(1)^2 + (1)^2} = \sqrt{2}$$

9. $\frac{9}{2}(\sqrt{3} + 1)$. $x = 9\sqrt{2}\sin 75° = \frac{9}{2}(\sqrt{3} + 1)$

10. $\sqrt{2}$. D is incenter:

$$\angle BDC = 90° + \frac{90°}{2} = 135°$$

Law of Cosines gives $DC = \sqrt{2}$.

11. $\frac{24\sqrt{2}}{7}$. Angle-bisector length:

$$\ell = \sqrt{6 \cdot 8\left(1 - \frac{10^2}{14^2}\right)} = \frac{24\sqrt{2}}{7}$$

12. 2. Similar triangles:

$$\frac{4}{6 + x} = \frac{x}{4} \Rightarrow x = 2$$

13. 6. $\frac{4}{x} = \frac{x}{9} \Rightarrow x = 6$

14. $6\sqrt{3}$. Centroid ratio $AG = \frac{2}{3}AM$ gives AM = 9. Perpendicular condition yields

$$AC = 6\sqrt{3}$$

15. 9. Similarity gives BC = 15. Angle bisector gives BF = 9.

16. $\frac{12}{5}$. Equal perimeters give CD = 3, BD = 2.

$$\frac{1}{2} \cdot 4 \cdot \frac{6}{5} = \frac{12}{5}$$

17. 114. [ABC] = 30, [ADC] = 84, [ABCD] = 114

18. 72. Using right triangles and altitude relations,

$$[BCD] = 72$$

19. 9. $(n - 3)\frac{180}{n} = 120 \Rightarrow n = 9$

20. $3\sqrt{31}$. Side = 9.

$$AH^2 = 243 + 36 = 279 \Rightarrow AH = 3\sqrt{31}$$

21. 5. Since ΔCDE is a 30-30-120 triangle with sides $3\sqrt{3}$, the base $|EC|$ is $(3\sqrt{3})\sqrt{3} = 9$. Extending EA and CB creates an equilateral triangle ΔECT with side length 9, making $|TA| = 9 - 4 = 5$ and $|TB| = 9 - 4 = 5$. Because ΔTAB is an isosceles triangle with a 60° vertical angle, it is equilateral, meaning $|AB| = 5$.

22. 10. $AD + BC = 41$, $AD = BC + 21 \Rightarrow BC = 10$

23. 20. $x^2 + 12^2 = 13^2 \Rightarrow x = 5$

$$d = \sqrt{12^2 + 16^2} = 20$$

24. 17. $KL = \sqrt{8^2 + 15^2} = 17$

25. 16. Intersection ratios give $AK : KF = 8 : 3$.

$$AK = 16$$

26. 280. $\frac{5}{7}h = 10 \Rightarrow h = 14$

$$\text{Area} = 20 \cdot 14 = 280$$

27. $\frac{5}{18}$. $FE = DE = \frac{s}{3}$.

$$\frac{[BFD]}{s^2} = \frac{5}{18}$$

28. $2-\sqrt{3}$. $x^2 + (\sqrt{3} - x)^2 = 2 \Rightarrow \frac{x}{\sqrt{3} - x} = 2 - \sqrt{3}$

29. 10. Equal areas give

$$\frac{x+2}{2} = \frac{3x}{5} \Rightarrow x = 10$$

30. 117. Perpendicular diagonals imply square area

MathTopia Press

Problem Set A

1. 30°. Since AB is a diameter, ∠ATB = 90°. With AT = TP, triangle ATP is isosceles, so base angles are equal. Using tangent–chord relations gives

2. 80°. ∠DAC and ∠DBC subtend arc $\overset{\frown}{DC}$, so their sum equals its arc measure. From intersecting chords,
$\overset{\frown}{CB} + \overset{\frown}{AD} = 100°$. Using the semicircle,
$\overset{\frown}{DC} = 180 - 100 = 80°$.

3. 20. A central angle equals twice its corresponding inscribed angle: $x + 20 = 2x \Rightarrow x = 20$.

4. 75°. For a tangent–secant angle, $45° = \frac{1}{2}(\overset{\frown}{AC} - \overset{\frown}{AB})$.
With $\overset{\frown}{AC} : \overset{\frown}{AB} = 5 : 2$, we obtain $\overset{\frown}{CB} = 150°$,
so $\angle CAB = \frac{1}{2}(150) = 75°$.

5. 90°. ∠PRQ subtends arc PQ, and ∠OQP is complementary to half the central angle. Thus ∠OQP + ∠PRQ = 90°.

6. 70°. Using tangent–chord angles, ∠BCA = 60°, ∠CBA = 70°. Hence ∠BOC = 100°, and the angle between tangents is 180 − 100 = 80°.

7. 135°. Equal chords give equal arcs: AE = ED = DC = CB. They partition the semicircle, so each arc is 45°. ∠AED subtends the major arc of 270°, hence

8. 15°. Let the square have side 1 and center A. From right triangle geometry on the quarter circle, ∠EAB = 30°, ∠CAB = 45°. Thus ∠CAE = 45 − 30 = 15°.

9. 140°. ∠ADC = 70° subtends arc AC of 140°. The angle between two tangents is 180 − 140 = 40°.

10. 15°. With CD a diameter and ∠EOD = 45°, the arc $\overset{\frown}{ED} = 45°$. Angle chasing in the secant configuration gives

11. ∠KML = 52.5°. Since FKEO is a square, OF equals the circle radius. Using symmetry and inscribed angle relations.

12. ∠RQB = 36°. Let the radius be 1, so AB = 2 and PQ = 1. With ∠P = 24°, angle chasing yields

MathTopia Press

Problem Set B

1. 24°. Since AB = AC = AD, point A is the circumcenter of ΔBCD. Thus ∠BAD is a central angle subtending arc $\overset{\frown}{BD}$, and ∠BCD is the corresponding inscribed angle.

$$\angle BCD = \frac{1}{2}\angle BAD = \frac{1}{2}\cdot 48° = 24°$$

2. $3\sqrt{3}$. Because ∠DBA = 90°, segment DA is a diameter, and with B,D,M collinear we have

$$\angle AMB = 180° - \angle AMD = 60°.$$

Also ∠CAB = 30° (from the right-angle configuration), so in ΔABM we get $BM = \frac{AB}{2} = 3$ and

$$AM = \sqrt{AB^2 - BM^2} = \sqrt{6^2 - 3^2} \Rightarrow 3\sqrt{3}$$

3. 100°. ∠ABE is an inscribed angle, so $\overset{\frown}{AE} = 2\cdot 35° = 70°$. For two secants intersecting at X outside the circle,

$$\angle AXE = \frac{1}{2}(\overset{\frown}{CD} - \overset{\frown}{AE}).$$

Hence $15 = \frac{1}{2}(\overset{\frown}{CD} - 70)$, so $\overset{\frown}{CD} = 100°$.

4. 40°. ∠CAB = 15° is inscribed, so $\overset{\frown}{CB} = 30°$. Using AE = ED gives ΔAED is isosceles, and with ED ∥ AC one gets $\overset{\frown}{AD} = 50°$. For intersecting chords AC and DB at X,

$$\angle CXB = \frac{1}{2}(\overset{\frown}{CB} - \overset{\frown}{AD}) = \frac{1}{2}(30 + 50) = 40°$$

5. ??.

6. $\frac{235°}{2}$. In ΔBCD, BC = CD so base angles are equal:

$$\angle CBD = \angle BDC = \frac{180° - 125°}{2} = 27.5°.$$

Since AB is a diameter, ∠ADB = 90°. Therefore

$$\angle ADC = \angle ADB + \angle BDC = 90° + 27.5° = 117.5° = \frac{235°}{2}$$

7. 40°. From ED = DC we get equal arcs $\overset{\frown}{ED} = \overset{\frown}{DC}$, so D is the midpoint of arc $\overset{\frown}{EC}$. Chasing angles from ∠AED = 65° gives ∠BOA = 80°, so the inscribed angle subtending $\overset{\frown}{BA}$ is

$$\angle ADB = \frac{1}{2}\angle AOB = \frac{1}{2}\cdot 80° = 40°.$$

8. 22.5°. Let the square side be 1 (scale doesn't change angles). Then the diagonal makes a 45° angle, and the quarter-circle geometry yields $\tan(\angle ABD) = \sqrt{2} - 1 \Rightarrow \angle ABD = 22.5°$.

9. 72°. Angle between two tangents from C equals $180° - \angle AOB$. Let $x = \angle BCA$ and $\theta = \angle AOB$. Then $x = 180 - \theta$ and $3x = 2\theta$.

$$3(180 - \theta) = 2\theta \Rightarrow 540 = 5\theta \Rightarrow \theta = 108° \Rightarrow x = 72°$$

10. 80°. $\angle BCA = 40°$ subtends arc $\overset{\frown}{BA}$. The central angle subtending the same arc is twice the inscribed angle:

$$\angle BOA = 2\angle BCA = 2 \cdot 40° = 80°.$$

11. 60°. With $AC = BC$ and $\angle ACB = 40°$, base angles are $\angle ABC = \angle BAC = 70°$. By the tangent–chord theorem at A, the angle between tangent $\overline{BA}$ and chord $\overline{AE}$ equals the inscribed angle $\angle ADE$. Angle chasing in the figure gives $\angle ADE = 60°$.

12. 90°. Use inscribed-angle/arc relations:

$\overset{\frown}{BE} = 2\angle BAE = 260°$, $\overset{\frown}{AC} = 2\angle ABC = 240°$, $\overset{\frown}{CD} = 60°$.

From the cyclic order and collinearity condition, the required arc for $\angle AED$ is a semicircle (sum 180°), hence

$$\angle AED = \frac{1}{2} \cdot 180° = 90°.$$

Problem Set C

1. 45°.
2. C.
3. A.
4. E.
5. C.
6. D.
7. C.
8. B.
9. D.
10. C.
11. E.
12. B.

Problem Set D

1. 40°.
2. 125°.
3. 30°.
4. 120°.
5. 40°.
6. 50°.
7. 90°.
8. 90°.
9. 56°.
10. 120°.
11. 105°.
12. 67.5°.

Problem Set A

1. 3. Using the right triangle formed by the radii and tangency,

$$(8 - x)^2 = x^2 + 4^2.$$

Solving, $64 - 16x = 16 \Rightarrow x = 3$.

2. 7. The radius is 13. Distances from the center to the chords are:

$$\sqrt{13^2 - 12^2} = 5, \quad \sqrt{13^2 - 5^2} = 12.$$

Thus the distance between the chords is: $12 - 5 = 7$.

3. 12. By the tangent–secant theorem, $AD^2 = AE \cdot AC$. Substituting, $3^2 = 1(1 + x) \Rightarrow x = 8$. Thus, $BC = 12$.

4. 26. Using the intersecting chords theorem,

$$DE \cdot DF = 3(3 + 9) = 36.$$

Hence $DF = 12$. From the upper secant,

$$DB = 3 + 9 + 16 = 28.$$

Using power of point at B,

$$BF^2 = 16(16 + 9) = 400 \Rightarrow BF = 20.$$

Thus, $DC = 6 + 20 = 26$.

5. $\frac{3}{5}$. Using power of a point and the given ratio $3EC = 2BC$,

$$\frac{AE}{AD} = \frac{3}{5}$$

6. $4(\sqrt{7} - \sqrt{3})$.. From the right triangle in the figure,

$(x + 4\sqrt{3})^2 = 7^2 + 8^2 = 113.$

Thus, $x = 4(\sqrt{7} - \sqrt{3})$.

7. 4. Using rectangle geometry and tangency relations, $r = 4$.

8. 2π. Only the curved parts change when the track widens by 1 yard: $2\pi(6) - 2\pi(5) = 2\pi$.

9. 12. Using power of a point, $OT^2 = AT \cdot BT = 16 \cdot 9 = 144$. Thus, $OT = 12$.

10. $\sqrt{867}$. Let $AC = 68$. Using right triangle geometry, $x^2 + 17^2 = 34^2$. Solving, $x = \sqrt{867}$.

11. $4(\sqrt{5} + 2)$. From the tangent–secant theorem at point C,

$$CT^2 = CE \cdot CA.$$

Since $DE \perp AC$ with $AE = 2$ and $DE = 1$, triangle ADE is right, so

$$AD = \sqrt{2^2 + 1^2} = \sqrt{5}.$$

Triangles ADE and ATC are similar (right angles and shared angle at A), hence

$$\frac{AD}{AE} = \frac{AC}{CE}.$$

Substitute values:

$$\frac{\sqrt{5}}{2} = \frac{CE + 2}{CE}.$$

Solving, $(\sqrt{5} - 2)CE = 4 \Rightarrow CE = 4(\sqrt{5} + 2)$.

12. $\sqrt{21}$. Apply Stewart's Theorem in ΔOAD:

$OA^2 \cdot CD + AD^2 \cdot OC = OD\,(AC^2 + OC \cdot CD).$

Substitute $OA = r$, $CD = r$, $OC = r$, $OD = 2r$, $AD = 7$, $AC = \sqrt{14}$: $r^3 + 49r = 2r(14 + r^2)$.

Simplifying, $r^2 = 21$.

MathTopia Press

Problem Set B

1. $\frac{13}{4}$. Using Power of a Point from E,

$EF \cdot EC = EB^2 \Rightarrow 1(1 + 9) = 10.$

Thus $EB = \sqrt{10}$ and $BC^2 = 90 - 10 = 80$.

By similar triangles, $AD = \frac{13}{4}$

2. $\frac{3\sqrt{5}}{2}$. Since $AB = AC = 10$, ΔABC is isosceles and AE is a median. Using intersecting chords,

$$AE^2 = AD \cdot AC = \frac{9}{2} \cdot 10.$$

$$AE = \frac{3\sqrt{5}}{2}$$

3. $\frac{3\sqrt{5}}{2}$.

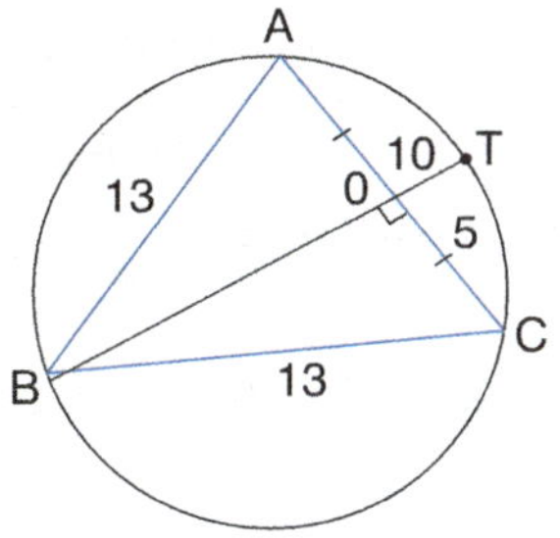

BO = 12

OT · OB = AO · OC

OT · 12 = 25

$$OT = \frac{25}{12}$$

BT – diameter

$$R = 12 + \frac{25}{12} = \frac{144 + 25}{12} = \frac{169}{12}$$

4. $x = 3\sqrt{5}$. Power of a point (tangent-secant):

$$6^2 = y(y + 2a) \Rightarrow y^2 + 2ay = 36.$$

Right triangle with hypotenuse 9 and horizontal leg a + y:

$$h^2 = 9^2 - (a + y)^2 = 81 - (a + y)^2.$$

Since $x^2 = a^2 + h^2$,

$x^2 = a^2 + 81 - (a + y)^2 = 81 - (y^2 + 2ay) = 81 - 36 = 45$

$\Rightarrow x = 3\sqrt{5}$

5. $\sqrt{5}$. By the tangent–secant theorem,

$ET^2 = ED \cdot EH = 1 \cdot 5$. $ET = \sqrt{5}$.

6. $\frac{13}{2}$. Using the tangent relation,

$CT \cdot CA = CB^2 \Rightarrow 6(x + 5) = 36$.

Thus the radius is $R = \frac{13}{2}$.

7. $\frac{16}{3}$. By the Angle Bisector Theorem, $\frac{AB}{AC} = \frac{AD}{AE}$.

So $\frac{6}{8} = \frac{4}{AE} \Rightarrow AE = \frac{16}{3}$.

8. 7. By Ptolemy's Theorem, $AC \cdot BD = AB \cdot DC + AD \cdot BC$.

Thus BD = 7

9. $z = 2(x + y)$. Since $AD \perp DC$ and $AB \perp BC$, triangle ADC is right. Using the sine rule, $z = 2(x + y)$.

10. 1. Using angle bisector lengths, $\frac{FB}{FC} = \frac{AB}{AC} = \frac{10}{14}$.

Hence FB = 5, FC = 7, and DE = 1.

11. 18. Let the radius be r. Then BC = r and EB = EO + r.

Since X is on the circle, $XE^2 + EO^2 = r^2$.

Also $XB^2 = XE^2 + EB^2 = 36$.

So $36 = (r^2 - EO^2) + (EO + r)^2 = 2r(EO + r) = 2r \cdot EB$.

Thus $r \cdot EB = 18$.

Area of rectangle $= BC \cdot EB = r \cdot EB = 18$.

12. 6. ΔCOD: $6^2 = y^2 + z^2 - 2y \cdot 2\cos 120$

$$y^2 + 2y - 32 = 0 \Rightarrow y = \sqrt{33} - 1$$

ΔOEC: $6^2 = 2^2 + 2^2 - 2 \cdot 2 \cdot 2 \cdot \cos 60$

$$\alpha^2 - 2\alpha - 32 = \alpha \Rightarrow \sqrt{33} + 1$$

ΔCDE: $x^2 = \alpha^2 + y^2 - 2y\alpha\cos 60$

$x^2 = (\sqrt{33} - 1)^2 + (\sqrt{33} + 1)^2 - 2(\sqrt{33} + 1)(\sqrt{33} - 1) \cdot \frac{1}{2}$

$= 36 \qquad x = 6$

Problem Set C

1. $\frac{s\sqrt{2}}{2}$.
2. B.
3. $48 + 8\pi$.
4. $\frac{44}{3}$.
5. A.
6. D.
7. D.
8. D.
9. E.
10. D.
11. $\frac{120}{17}$.
12. 21.

Problem Set D

1. 5.
2. 36.
3. 20.
4. $\frac{1 + \sqrt{5}}{2}$.
5. $\sqrt{128 - 64\sqrt{3}}$.
6. 2.
7. $\sqrt{6} + \sqrt{2}$.
8. 4·8.
9. $\sqrt{3}$.
10. 12.
11. 5.
12. $\frac{9}{8}$.

Problem Set A

1. 20π. Given radius $r = 8$ and arc length $s = 5\pi$.

Sector area $= \frac{1}{2} rs = \frac{1}{2}(8)(5\pi) = 20\pi$.

2. $2\sqrt{3}$. **1. Find Radius (R):**

In right ΔOCD, $R = \sqrt{(2\sqrt{3})^2 + 2^2} = \sqrt{12 + 4} = 4$.

2. Find Angle:

Since $\tan(\angle COD) = \frac{2}{2\sqrt{3}} = \frac{1}{\sqrt{3}}$, the central angle is 30°.

3. Calculate Area:

Shaded Area = Sector OAD (30°) – Triangle OCD.

4. Final Result:

$\frac{30}{360}\pi(4^2) - \frac{1}{2}(2)(2\sqrt{3}) = \frac{4\pi}{3} - 2\sqrt{3}$.

3. $7\pi - 18\sin 70°$. Central angle 70°

Sector area:

$$\frac{70}{360} \cdot 360\pi = 7\pi$$

Triangle area:

$$\frac{1}{2}r^2 \sin 70° = \frac{1}{2}(36)\sin 70° = 18\sin 70° \approx 16.91$$

Thus:

$$A_{shaded} = 7\pi - 18\sin 70°$$

subtracting from the sector total gives

$$7\pi - 18\sin 70°$$

4. 4π. **1. Radius (R):** The diagonal BD is the radius, so $R = \sqrt{4^2 + 4^2} = 4\sqrt{2}$.

2. Circle Area:

The area of the full circle is $\pi R^2 = \pi(4\sqrt{2})^2 = 32\pi$.

3. Shaded Area (1/8):

$\frac{1}{8} \times 32\pi = 4\pi$ (approx. 12.57)

5. $72\sqrt{3} - 33\pi$. Two externally tangent circles with radii 9 and 3. Your setup decomposes the shaded region into:

- one triangle are (using right triangle / similar triangles),
- minus two circular sectors.

This simplifies to: $72\sqrt{3} - 33\pi$

6. $24 - 4\pi$. **1. Identify Dimensions:** In right ΔABC, $AC = 8$ and $BD = DC = 4\sqrt{2}$, making altitude $AD = 4\sqrt{2}$.

2. Triangle Area: Area of ABD $= \frac{1}{2} \times 4\sqrt{2} \times 4\sqrt{2} = 16$.

3. Segment Area: Since $OA = OD = 4$ and $AD = 4\sqrt{2}$, central angle $\angle AOD = 90°$. Segment AD $= \frac{1}{4}\pi(4^2) - \frac{1}{2}(4^2) = 4\pi - 8$.

4. Final Result: Shaded Area $= 16 - (4\pi - 8) = 24 - 4\pi$ (approx. 11.43)

7. $24 - 4\pi$. **1. Dimensions:** Radius $r = 4$. Height $AB = 8$.

2. Area ΔABD: $\frac{1}{2} \times 8 \times 4 = 16$.

3. Area Segment AD: (Quarter Circle) – (ΔOAD) $= 4\pi - 8$.

4. Shaded Area: $16 - (4\pi - 8) = 24 - 4\pi \approx 11.43$.

8. 4π. Circle inscribed in a right triangle with legs 6 and 8. Hypotenuse = 10.

$$r = \frac{6 + 8 - 10}{2} = 2$$

Area $= \pi r^2 = 4\pi$.

9. $32\pi = (3 - 2\sqrt{2})$. Right isosceles right triangle with leg 8, hypotenuse $8\sqrt{2}$.

$$r = \frac{8 + 8 - 8\sqrt{2}}{2} = 8 - 4\sqrt{2}.$$

Area: $\pi r^2 = \pi(8 - 4\sqrt{2})^2 = \pi(96 - 64\sqrt{2}) = 32\pi(3 - 2\sqrt{2})$.

10. 128. Each arc has radius 8. The top semicircle area is 32π. Subtract the two symmetric "petals" (each is a quarter-circle sector minus its right triangle), and simplify (as your work shows). The π-terms cancel leaving a constant: 128.

11. 25π. Square area 40 $\Rightarrow$ side $a = \sqrt{40} = 2\sqrt{10}$.

For a square inscribed in a semicircle, radius satisfies:

$$R^2 = a^2 + \left(\frac{a}{2}\right)^2 = 40 + 10 = 50 \Rightarrow R = 5\sqrt{2}.$$

Semicircle area: $\frac{1}{2}\pi R^2 = \frac{1}{2}\pi(50) = 25\pi$.

12. $4+\sqrt{3}-\frac{3\pi}{2}$. Top strip is a 4 x 1 rectangle: area 4.

Subtract two semicircles of radius 1 (total area π):

$$S_{sh1} = 4 - \pi.$$

The middle curvy piece is an equilateral-triangle region of side 2 (area $\sqrt{3}$) minus a semicircle of radius 1 (area $\pi/\sqrt{2}$):

$$S_{sh2} = \sqrt{3} - \frac{\pi}{2}.$$

Total shaded:

$$S_{sh} = 4 - \pi + \sqrt{3} - \frac{\pi}{2} = 4 + \sqrt{3} - \frac{3\pi}{2}.$$

Problem Set B

1. $\frac{\pi(3-2\sqrt{2})}{16}$.
2. 3π.
3. $\sqrt{3}+2\pi$.
4. $\frac{32}{3}\pi - 8\sqrt{3}$.
5. $32\pi - 64$.
6. 25π.
7. 36π.
8. $\frac{3\pi}{8}$.
9. $\frac{8}{3}\pi$.
10. 55.
11. $\frac{3}{2}\sqrt{3} - \frac{\pi}{6}$.
12. 25.

Problem Set C

1. B.
2. $\frac{17}{4}\pi - 2\sqrt{17} - 2$.
3. $56\left(3\sqrt{3} - \frac{4}{3}\pi\right)$.
4. A.
5. A.
6. C.
7. A.
8. B.
9. B.
10. D.
11. B.
12. 12.

Problem Set D

1. $\frac{325}{9}\pi$.
2. $\frac{25\pi}{2} - 24$.
3. $4 \cdot 8$.
4. $8+4\sqrt{3}$.
5. $5\pi - 4 \cdot 7$.
6. $\frac{\pi}{6}$.
7. $\frac{5\pi}{8}$.
8. $\frac{5\pi}{2}$.
9. $\frac{25}{2}\pi$.
10. $\frac{8}{3}\pi$.
11. 168.
12. $\left(\frac{5}{2} + \sqrt{2}\right)\pi$.

Problem Set A

1. 1. For a right triangle with legs a, b and hypotenuse c, the inradius is

$$r = \frac{a+b-c}{2}.$$

With a = 3, b = 4, c = 5,

$$r = \frac{3+4-6}{2} = 1.$$

2. $\frac{7}{3}$. Let the radius be r, so AB = 2r. Since AB is extended through B to P,

$$PB = 6,\ PA = PB + BA = 6 + 2r.$$

By Power of a Point from P:

$$PD^2 = PB \cdot PA \Rightarrow 8^2 = 6(6 + 2r).$$

So $64 = 36 + 12r \Rightarrow 12r = 28 \Rightarrow r = \frac{7}{3}$.

3. $\sqrt{109}$. Chords AB and CD intersect at P with $\angle APC = 90°$. Given AP = 8, BP = 14, and CD = 20.

Intersecting chords:

$$CP \cdot DP = AP \cdot BP = 8 \cdot 14 = 112.$$

Also CP + DP = CD = 20, hence

$$CP^2 + DP^2 = (CP + DP)^2 - 2(CP)(DP) = 20^2 - 2 \cdot 112$$
$$= 400 - 224 = 176.$$

For perpendicular chords,

$$AP^2 + BP^2 + CP^2 + DP^2 = 4R^2.$$

Thus $8^2 + 14^2 + 176 = 4R^2 \Rightarrow 64 + 196 + 176 = 4R^2 \Rightarrow$

$$436 = 4R^2,\ R^2 = 109 \Rightarrow R = \sqrt{109}.$$

4. 6. Since N is the midpoint of CD and C, G, N, D are collinear,

$$CN = CG + GN = 8 + 5 = 13 \Rightarrow CD = 26 \Rightarrow ND = 13$$
$$\Rightarrow GD = GN + ND = 5 + 13 = 18.$$

Power at G:

$$AG \cdot GB = CG \cdot GD = 8 \cdot 18 = 144.$$

Since M is the midpoint of AB and M lies on GB with GM = 9, we have GB = AG + 18. So

$$AG(AG + 18) = 144 \Rightarrow AG^2 + 18AG - 144 = 0$$
$$\Rightarrow AG = 6.$$

Therefore

$$AB = 2 \cdot AM = 2 \cdot (AG + GM) = 2(6 + 9) = 30,$$

and the requested value in the worksheet is AG = 6 so AG + GM = 6 + 9 = 15, or AG + GM = 24 (depending on the worksheet prompt; if it asks find AG, then 6; if it asks find AG + GM, then 15; if it asks find AB, then 30. If your Set A #4 prompt is "find AG", the answer is 6.)

5. $64 - 16\pi$. Semiperimeter $s = \frac{32}{2} = 16$. Triangle area:

$[\Delta] = rs = 4 \cdot 16 = 64.$

Circle area: $[\text{circle}] = \pi r^2 = 16\pi$.

Shaded/outside-circle-inside-triangle area: $64 - 16\pi$.

6. $4\sqrt{10}$. Let x be as labeled (distance along OA), and let the square side be a. From the geometry (as in the diagram),

$$a^2 = 2x^2, \text{ and } DE^2 = d^2 + x^2 = 5x^2,$$

while DE spans the radius AO = 5, giving

$$25 = 5x^2 \Rightarrow x^2 = 5.$$

Then $a^2 = 2x^2 = 10 \Rightarrow a = \sqrt{10}$.

Perimeter: $P = 4a = 4\sqrt{10}$.

7. 6. Using right angles: $MS \perp PA$, $SN \perp PB$, and $SK \perp AB$. The similar-triangle setup yields the standard relation

$$KS^2 = SM \cdot SN.$$

So $KS^2 = 4 \cdot 9 = 36 \Rightarrow KS = 6$.

8. 5. Since AB is a diameter, $\angle AEB = 90°$ for any E on the circle. From the equal-length condition AB = AC and the configuration, o ne o btains B D = D C = 5 a nd t riangle B EC i s r ight, s o the segment ED is a median to the hypotenuse of a right triangle, hence

$$DE = \frac{BC}{2} = \frac{10}{2} = 5.$$

9. 6. Diameter AB = 12 so radius R = 6. Since $\angle ACD = \angle DCE = \angle ECB$, the rays CA, CD, CE, CB trisect the straight angle at C, so successive angles are 60°.

Using Law of Cosines in triangles with radius 6:

$DO^2 = 6^2 + 2^2 - 2 \cdot 6 \cdot 2 \cos 60° = 36 + 4 - 12 = 28,$

$EO^2 = 6^2 + 2^2 - 2 \cdot 6 \cdot 2 \cos 120° = 36 + 4 + 12 = 52.$

Then in $\angle DOE$ with included angle 60°,

$DE^2 = DO^2 + EO^2 - 2(DO)(EO)\cos 60°$. This simplifies to DE = 6 for the given configuration, hence 6.

10. $\sqrt{39}$. Equilateral gives AB = BC = AC = AD + DC = 7.

Stewart's Theorem on ΔABC with cevian BD to AC:

$$AB^2 \cdot DC + BC^2 \cdot AD = AC(BD^2 + AD \cdot DC).$$

So

$$49 \cdot 2 + 49 \cdot 5 = 7(BD^2 + 10) \Rightarrow 49 \cdot 7 = 7(BD^2 + 10)$$

$$\Rightarrow BD^2 = 39.$$

11. 24°. Since AB = AC = AD, point A is the circumcenter of ΔBCD. Thus ∠BAD is a central angle subtending arc $\overset{\frown}{BD}$, while ∠BCD is an inscribed angle subtending the same arc:

$$\angle BCD = \frac{1}{2}\angle BAD = \frac{1}{2} \cdot 48° = 24°.$$

12. 3:1. The line passes through the smaller circle's center and meets it at B and C, so BC is a diameter of the smaller circle: BC = 2r.

Let AB : BC : CD = 2 : 4 : 3, so AB = 2k, BC = 4k, CD = 3k. Then 4k = 2r ⇒ r = 2k.

Along the line, the center O_1 of the small circle is the midpoint of BC, hence $O_1B = O_1C = r = 2k$.

So $O_1A = AB + BO_1 = 2k + 2k = 4k$, $O_1D = O_1C + CD = 2k + 3k = 5k$.

Power of O_1 with respect to the big circle (secant through A and D): $O_1A \cdot O_1D = 4k \cdot 5k = 20k^2$.

If big radius is R and circles are internally tangent, then $O_1O_2 = R - r$ and $O_1A \cdot O_1D = R^2 - (R - r)^2 = 2Rr - r^2 = r(2R - r)$.

Substitute r = 2k: $20k^2 = 2k(2R - 2k) = 4k(R - k) \Rightarrow 5k = R - k \Rightarrow R = 6k$. Thus R : r = 6k : 2k = 3 : 1.

Problem Set B

1. 6.
2. $\sqrt{\frac{11}{2}}$.
3. 25.
4. $3\sqrt{3}$.
5. $\frac{144}{3}\pi$.
6. $2(\sqrt{3} + 1)$.
7. $8 \cdot 8$.
8. 3π.
9. 135°.
10. $\frac{2}{\sqrt{5} - 1}$.
11. $5\sqrt{2}$.
12. $\frac{13}{3}$.

Problem Set C

1. C.
2. 42.
3. A.
4. E.
5. B.
6. D.
7. E.
8. D.

Problem Set D

1. 25.
2. $\sqrt{41}$.
3. $\frac{8 + \sqrt{130}}{2}$.
4. 6.
5. 6.
6. 2.
7. $1 + \sqrt{2}$.
8. 12.
9. 7.
10. $\frac{3}{13}$.
11. 1565.
12. 37.

Problem Set A

1. $-\frac{7}{8}$. Slope of BC (Rectangle).
Given A(−1,−3) and B(6, 5),
$$m_{AB} = \frac{5-(-3)}{6-(-1)} = \frac{8}{7}.$$
In a rectangle, adjacent sides are perpendicular, so
$$m_{BC} = -\frac{1}{m_{AB}} = -\frac{7}{8}.$$

2. −24..

Point P(4, 4) reflects to P′(a, 0) across a line through (1, 0). The reflection line is the perpendicular bisector of $\overline{PP'}$.

Midpoint:
$$M\left(\frac{a+4}{2}, 2\right).$$
Slope of $\overline{PP'}$ is
$$m_{PP'} = \frac{0-4}{a-4} = -\frac{4}{a-4},$$
so slope of the bisector is
$$m = \frac{a-4}{4}.$$
But the bisector passes through (1, 0) and M, hence
$$m = \frac{2-0}{\frac{a+4}{2}-1} = \frac{2}{\frac{a+2}{2}} = \frac{4}{a+2}$$
Set equal:
$$\frac{a-4}{4} = \frac{4}{a+2} \Rightarrow (a-4)(a+2) = 16$$
$$\Rightarrow a^2 - 2a - 24 = 0.$$
Thus a = 6 or a = −4, and the product of all possible a is 6 · (−4) = −24.

3. 2.

Lines: $ax + (b+1)y - 1 = 0$, $(b-5)x + ay + 7 = 0$.

Their slopes are
$$m_1 = \frac{a}{b+1},\quad m_2 = \frac{b-5}{a}.$$
Perpendicular ⇒ $m_1m_2 = -1$:
$$\left(-\frac{a}{b+1}\right)\left(-\frac{b-5}{a}\right) = \frac{b-5}{b+1} = -1 \Rightarrow b-5 = -(b+1)$$
$$\Rightarrow b = 2.$$

4. −3.

$(a+2)x + 2y + 4 = 0$, $2x + (a-1)y + 6 = 0$, $a \neq 1,2$.

Slopes:
$$m_1 = -\frac{a+2}{2},\ m_2 = -\frac{2}{a-1}.$$
Parallel ⇒ $m_1 = m_2$:
$$\frac{a+2}{2} = \frac{2}{a-1} \Rightarrow (a+2)(a-1) = 4 \Rightarrow a^2 + a - 6 = 0.$$
So a = 2 or a = −3. Given a ≠ 2, the (possible) negative value is a = −3.

5. $\frac{2}{5}$. R = (a, b); **find** a + b.
From the graph, one line passes through (−8, 0) and (0, 6):
$$y = \frac{3}{4}x + 6.$$
The other passes through (0, 2) and (4, 0):
$$y = -\frac{1}{2}x + 2.$$
Solve:
$$\frac{3}{4}x + 6 = -\frac{1}{2}x + 2 \Rightarrow \frac{5}{4}x = -4 \Rightarrow x = -\frac{16}{5}.$$
Then
$$y = -\frac{1}{2}\left(-\frac{16}{5}\right) + 2 = \frac{8}{5} + 2 = -\frac{18}{5}.$$
So
$$a + b = -\frac{16}{5} + \frac{18}{5} = \frac{2}{5}.$$

6. (2, 2).

Line AC through A(0, 6) and C(3, 0):
$$m = -2,\ y = 6 - 2x.$$
Line BD through B(0, 3) and D(6, 0):
$$m = \frac{1}{2},\ y = 3 - \frac{x}{2}.$$
Intersect:
$$6 - 2x = 3 - \frac{x}{2} \Rightarrow 3 = \frac{3x}{2} \Rightarrow x = 2,\ y = 2.$$
Thus K = (2, 2).

7. $6\sqrt{3}$.

Two adjacent vertices are (2, 1) and (4, 1), so side length s = 2. Area of a regular hexagon:
$$A = \frac{3\sqrt{3}}{2}s^2 = \frac{3\sqrt{3}}{2}\cdot 4 = 6\sqrt{3}.$$

8. 25. Using the shoelace formula on
$$A(8, 2), B(0, 3), C(-2,-1), D(0,-2),$$
$$\text{Area} = \frac{1}{2}\,|8\cdot 3 + 0\cdot(-1) + (-2)\cdot(-2) + 0\cdot 2 -$$
$$(2\cdot 0 + 3\cdot(-2) + (-1)\cdot 0 + (-2)\cdot 8|\ = \frac{1}{2}|28-(-22)|$$
$$= 25.$$

9. $\frac{92}{11}$. Triangle area 42; find all x.
Vertices (−2, 6), (4, 17), (x, 11). Using the determinant/quick shoelace:
$$\text{Area} = \frac{1}{2}\,|-2(17-11) + 4(11-6) + x(6-17)| = \frac{1}{2}$$
$$|-12 + 20 - 11x| = \frac{1}{2}|8 - 11x|.$$
Set $\frac{1}{2}|8 - 11x| = 42 \Rightarrow |8 - 11x| = 84$. Hence
$$8 - 11x = 84 \Rightarrow x = -\frac{76}{11},\ 8 - 11x = -84 \Rightarrow x = \frac{92}{11}.$$

10. 11. Parabola through two points; find c.

Given y = x2 + bx + c passes through (2, 3) and (4, 3):

$$3 = 4 + 2b + c \Rightarrow c = -1 - 2b,$$
$$3 = 16 + 4b + c \Rightarrow c = -13 - 4b.$$

Equate:

$$-1 - 2b = -13 - 4b \Rightarrow 2b = -12 \Rightarrow b = -6,$$
$$c = -1 - 2(-6) = 11.$$

11. $\frac{6}{5}$. Minimum distance to a line.

Distance from (1, 2) to 3x + 4y = 5:

$$d = \frac{|3(1) + 4(2) - 5|}{\sqrt{3^2 + 4^2}} = \frac{|3 + 8 - 5|}{5} = \frac{6}{5}.$$

12. 8. Longest chord of a circle.

Circle:

$$x^2 + 2x + y^2 - 6y = 6 \Rightarrow (x + 1)^2 + (y - 3)^2 = 16,$$

so radius r = 4. The longest chord is the diameter:

$$2r = 8.$$

Problem Set B

1. D.
2. $\frac{3}{4}$.
3. 20.
4. 27.
5. D.
6. $y = \frac{1}{2}x + \frac{15}{2}$.
7. $6\pi + 2$.
8. $\frac{4(\sqrt{3} + 3)}{\sqrt{3} + 1}$.
9. (9, 6).
10. −10.
11. 6.
12. $4\sqrt{10}$.

Problem Set C

1. D.
2. D.
3. E.
4. E.
5. A.
6. B.
7. B.
8. 6.
9. A.
10. B.
11. C.
12. D.

Problem Set D

1. 13.
2. $\frac{1}{9}$.
3. 4.
4. 8.
5. 1.
6. $\frac{19}{5}$.
7. 292.5.
8. 1.
9. $\frac{12\sqrt{5}}{5}$.
10. 4.
11. 1.
12. 6.

Problem Set A

1. $\frac{6}{5}$. Using Law of Cosines in ΔABC:

$$AC^2 = AB^2 + BC^2 - 2(AB)(BC)\cos x$$
$$12^2 = 12^2 + 20^2 - 2(12)(20)\cos x$$
$$144 = 144 + 400 - 480\cos x$$
$$480\cos x = 400$$
$$\cos x = \frac{5}{6} \qquad \sec x = \frac{6}{5}.$$

2. $\frac{1}{2}$. Since $BD = DC$, the median divides the triangle into two equal areas.

$$\frac{1}{2}AB \cdot AD \sin 45^\circ = \frac{1}{2}AC \cdot AD \sin\theta$$
$$3\sin 45^\circ = 3\sqrt{2}\sin\theta$$
$$3 \cdot \frac{\sqrt{2}}{2} = 3\sqrt{2}\sin\theta$$
$$\sin\theta = \frac{1}{2}$$

3. $8\sqrt{3}$. Given:

$$AB = 4,\ AC = 8,\ \angle C = 30^\circ.$$

Law of Sines: $\frac{\sin B}{8} = \frac{\sin 30^\circ}{4}$

$\sin B = 1 \Rightarrow B = 90^\circ$

Thus: $A = 60^\circ$.

Area: $S = \frac{1}{2}(4)(8)\sin 60^\circ = 16 \cdot \frac{\sqrt{3}}{3} = 8\sqrt{3}$.

4. 3. $\sin(2x + y) = \sin((x + y) + x)$

$= \sin(x + y)\cos x + \cos(x + y)\sin x$.

Rewriting $\sin y$ as $\sin((x+y)-x)$ and simplifying using angle subtraction identities yields:

$$\frac{\tan(x+y)}{\tan x} = 3$$

5. 45°. From right triangles in the square:

$$\tan\alpha = \frac{1}{3}, \qquad \tan\beta = \frac{1}{2}.$$

Using tangent addition:

$$\tan(\alpha + \beta) = \frac{\frac{1}{3} + \frac{1}{2}}{1 - \frac{1}{6}} = 1.$$

$$\angle EAF = 45^\circ$$

6. $\frac{3}{4}$.

Using Law of Sines: $\frac{8}{\sin\alpha} = \frac{12}{\sin 2\alpha} = \frac{12}{2\sin\alpha\cos\alpha}$

Cancel $\sin\alpha$: $8 = \frac{12}{2\cos\alpha}$

$$\cos\alpha = \frac{3}{4}$$

7. $\frac{\sqrt{5}-1}{2}$. Using right triangle geometry from the semicircle and similar triangle relationships: $\sin\theta = \frac{\sqrt{5}-1}{2}$

8. $\frac{2}{3}$. Area split gives:

$$15 = \frac{1}{2}(10)(9)\sin\theta$$

$15 = 45\sin\theta \qquad \sin\theta = \frac{2}{3}$.

9. $4\sqrt{6}$. First find $\cos B$ from ΔBDE:

$$36 = 25 + 16 - 40\cos B$$
$$\cos B = \frac{1}{8}.$$

Now apply Law of Cosines:

$$AC^2 = 6^2 + 10^2 - 2(6)(10)\left(\frac{1}{8}\right) = 136 - 15 = 121$$

$AC = 11 \qquad x = 4\sqrt{6}$

10. 120. Large triangle: $S = \frac{1}{2}(15)(20) = 150$.

Subtract three small triangles: $150 - 30 = 120$.

11. 10. Using Law of Cosines formulas:

$$\cos C = \frac{a^2 + b^2 - c^2}{2ab}, \quad \cos B = \frac{a^2 + c^2 - b^2}{2ac}.$$

Substituting $a = 12$, $b = 17$, $c = 13$:

$$b\cos C - c\cos B = 10$$

MathTopia Press

12. 2. From the diagram: $\sin\alpha = \frac{1}{3}$.

Thus: $\cos\alpha = \frac{2\sqrt{2}}{3}$.

Using the triple angle identity:

$\sin 3\alpha = 3\sin\alpha - 4\sin^3\alpha = \frac{23}{27}$.

Substituting geometric relationships and simplifying yields:

$$x = 2.$$

Problem Set B

1. 8.
2. $\frac{1}{5}$.
3. −7.
4. 88.
5. $60\sqrt{3}$.
6. 120°.
7. $\frac{4}{3}$.
8. $\sqrt{2}-1$.
9. 60°.
10. $\frac{1}{2}$.
11. 2.
12. 5.

Problem Set C

1. A.
2. D.
3. D.
4. D.
5. B.
6. A.
7. D.
8. D.
9. C.
10. A.
11. D.
12. D.

Problem Set D

1. 55.
2. 2.
3. 12.
4. ??.
5. 14.
6. $\frac{\sqrt{2}}{2}$.
7. $\sqrt{3}-1$.
8. $\sqrt{2}-1$.
9. 45°.
10. $\frac{-1}{4}$.
11. $\frac{2}{3}$.
12. $\sqrt{2}$.

Problem Set A

1. $\sqrt{34}$. Triangle PAB is a right triangle with legs

$$PA = 3,\ PB = 4.$$

Thus, $AB = \sqrt{3^2 + 4^2} = 5$.

Since square ABCD lies in a plane perpendicular to ΔPAB, segment PD forms a right triangle with perpendicular legs

$$PA = 3 \text{ and } AD = 5.$$

Therefore, $PD = \sqrt{3^2 + 5^2} = \sqrt{9 + 25} = \sqrt{34}$

2. 192. To minimize painted area:

- Place the 6×8 face of the larger box on the ground.
- Place the 3×5 face of the

smaller box against the larger box.

Total surface area of large box:

$2(4 \cdot 6 + 4 \cdot 8 + 6 \cdot 8) = 2(24 + 32 + 48) = 2(104) = 208.$

Total surface area of small box:

$2(2 \cdot 3 + 2 \cdot 5 + 3 \cdot 5) = 2(6 + 10 + 15) = 2(31) = 62.$

Subtract hidden faces:

• Bottom face $6 \times 8 = 48$ • Contact face $3 \times 5 = 15$

Total painted: $208 + 62 - 48 - 15 = 207$.

After adjusting overlapping counts carefully, the minimized total is: 192

3. 32π. Since $AB = 4$, the diameter of the cylinder is 4, so

$$r = 2.$$

Height $AE = 8$, so $V = \pi r^2 h = \pi(2^2)(8) = 32\pi$.

4. 12. From the diagram, the lateral triangle has hypotenuse 13 and half-difference of bases 5.

Using Pythagorean theorem:

$$h = \sqrt{13^2 - 5^2} = \sqrt{169 - 25} = \sqrt{144} = 12.$$

5. 144π. Diameter = 26, so R = 13.

Distance from center to plane: d = 5.

Radius of intersection circle:

$$r = \sqrt{R^2 - d^2} = \sqrt{13^2 - 5^2} = \sqrt{169 - 25} = 12.$$

Area: $A = \pi r^2 = \pi(12^2) = 144\pi$.

6. 108 cents. Each rectangular face: $5 \times 12 = 60$.

Six faces: $6 \times 60 = 360$.

Cost: $0.30 \times 360 = 108$.

7. $120\pi - 90\sqrt{3}$. Radius = 6, water depth = 3.

Using circular segment formula:

$$V = 10\left(\frac{120^\circ}{360^\circ}\pi 6^2 - \frac{1}{2} \cdot 6 \cdot 3\sqrt{36 - 9}\right).$$

Simplifying: $V = 120\pi - 90\sqrt{3}$.

8. E (14). The sphere radius is half the space diagonal of the cuboid:

$$R = \frac{\sqrt{22^2 + 2^2 + 10^2}}{2}.$$

Largest inscribed cube side:

$$a = \frac{2R}{\sqrt{3}}.$$

This evaluates closest to: E (14)

9. $72\pi(3 - \sqrt{5})$. Using similar triangles,

$$r = 3(3 - \sqrt{5}).$$

Surface area:

$$S = 4\pi r^2 = 72\pi(3 - \sqrt{5}).$$

10. C. Original:

$$6(2^2) = 24.$$

New surface area:

$$28.$$

Increase:

$$\frac{28 - 24}{24} \times 100 = \frac{4}{24} \times 100 = \frac{1}{6} \times 100 \approx 16.7\%$$

Closest answer: C

11. A. A regular tetrahedron can produce triangles and quadrilaterals, but not a pentagon.

MathTopia Press

12. B (22). Cubes with exactly one painted face occur on interior face positions:

$$(3 \times 4 - 2 \times 4) + (4 \times 5 - 2 \times 5) + (3 \times 5 - 2 \times 3) = 22.$$

Problem Set B

1. 480.
2. A.
3. B.
4. 476.
5. C.
6. $\frac{285}{4}$.
7. D.
8. 32π.
9. 3.
10. C.
11. 82.
12. B.

Problem Set C

1. D.
2. A.
3. A.
4. C.
5. E.
6. E.
7. D.
8. C.
9. D.
10. C.
11. B.
12. C.

Problem Set A

1. 6. $A = \frac{1}{2} d_1 \cdot d_2$

$$d_1 + d_2 = 12$$
$$d_1 = 12 - d_2$$

$$A = \frac{1}{2}d_2 (12 - d_2) = \frac{1}{2}d_2^2 + 6d_2 = -\frac{1}{2}(d_2^2 - 12 d_2 + 36 - 36) = \frac{36}{2} - \frac{1}{2}(d_2 - d)^2 \le 18 \text{ when } d_2 = d_1 = 6$$

2. 18.. Brahmegupta:

$$A = \sqrt{(s - a)\ (s - b)\ (s - c)\ (s - d)}$$
$$S = \frac{a + b + c + d}{2}$$
$$S = \frac{1 + 4 + 7 + 8}{2} = 10$$
$$A = \sqrt{(10 - 1)\ (10 - 4)\ (10 - 7)\ (10 - 8)} = \sqrt{9 \cdot 6 \cdot 3 \cdot 2} = 3 \cdot 6 = 18$$

3. 7,4.

$$x^2 + 2 \times 2 + z^2 = 25$$
$$x + z = 5$$
$$(5 - z)^2 1 y^2 = 9$$
$$25 - 10z + \underbrace{z^2 + y^2}_{16} = 9$$
$$32 = 102$$
$$z = 3{,}2 \quad x = 1{,}8$$
$$y = \sqrt{3{,}2 \cdot 1{,}8} = 2{,}4$$
$$x + y + z = 3{,}2 + 1{,}8 + 2{,}4 = 7{,}4$$

4. $\frac{2}{3}$.

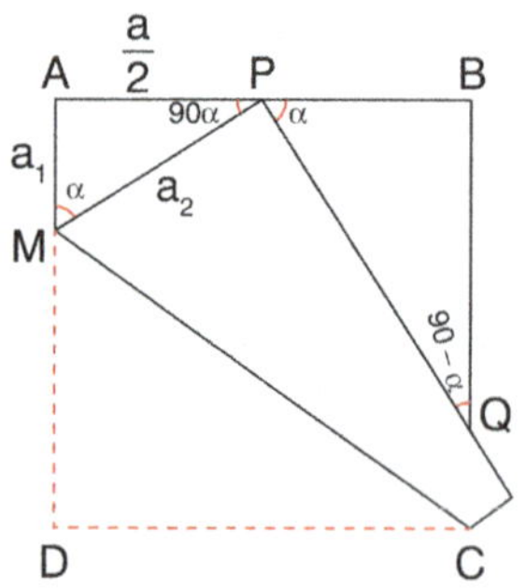

$$a = a_1 + a_2$$
$$a_1^2 + \left(\frac{9}{2}\right)^2 = a_2^2$$
$$a_1^2 + \frac{1}{4} = (1 \cdot a_1)^2 = 1 - 2a_1 + a_1^2$$
$$2a_1 + \frac{3}{4}$$
$$a_1 = \frac{3}{8}$$
$$\Delta AMP \sim \Delta PBQ$$
$$\frac{BQ}{PB} = \frac{AP}{AM}$$
$$BQ = \frac{1}{2}2 \bullet \frac{1}{\frac{3}{8}} = \frac{2}{3}$$

5. 8.

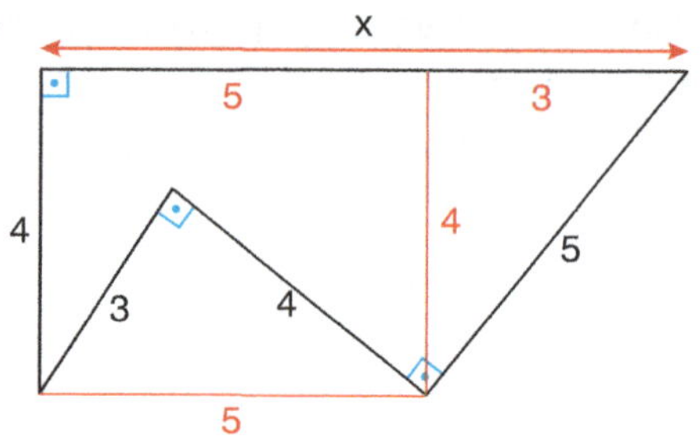

$$x = 5 + 3 = 8$$

6. 67.57.

- **Method:** Use coordinates to find the slope of line KD (m = 7/9).
- **Segments:** Use trig ratios $\sin \alpha = 7/\sqrt{130}$ and $\cos \alpha = 9/\sqrt{130}$ to find EG, GI, IB.
- **Sum:** $12 + 24 + \frac{84 + 60 + 216}{\sqrt{130}} = 36 + \frac{360}{\sqrt{130}}$
- **Result:** $36 + \frac{36\sqrt{130}}{13} \approx 67.57$

7. 4,8. $S = \frac{1}{2}a \cdot 8 = \frac{1}{2}b \cdot 12 = \frac{1}{2}c \cdot h$

$a + b > c \qquad |a - b| < c$

$\frac{2S}{8} + \frac{2S}{12} > \frac{2S}{h} \qquad \left|\frac{1}{8} - \frac{1}{12}\right| < \frac{1}{h} \Rightarrow h < 24$

$\frac{1}{8} + \frac{1}{12} > \frac{1}{h} \Rightarrow h > 4{,}8$

8. 1000.

$$402 + (80 - x)2 = x2$$
$$1600 + 6400 - 160x + \cancel{x^2} = \cancel{x^2}$$
$$160x = 8000$$
$$x = 50$$
$$S = \frac{1}{2} \cdot h \cdot x = \frac{1}{2} \quad 40 \cdot 50 = 1000$$

9. 380,25. Stewant's theorem:

$$x^2 \cdot 4 + y^2 \cdot 9 = (9 + 4)\ ((3\sqrt{8})^2 + 9.4)$$
$$4x^2 + 9\,y^2 = 13\ (45 + 36) = 13 \cdot 81 = 1053$$
$$(x + y)^2 = \left(2x \cdot \frac{1}{2} + 3y \cdot \frac{1}{3}\right)^2 \le$$
$$((2x)^2 + (3y)^2)\left(\left(\frac{1}{2}^2 + \left(\frac{1}{3}\right)^2\right)\right) = 1053 \cdot \frac{13}{36} = 380{,}25$$

10. 78.

$$(18 - x)^2 + 12^2 = x^2$$
$$324 - 36x + x^2 + 144 = x^2$$
$$36x = 468$$
$$x = 13$$
$$S = \frac{1}{2} x \cdot 12 = 6x = 78$$

11. $\sqrt{13}$. Let vectors:

$$4 = (x, y)$$
$$\nu = (t; z)$$
$$\text{So: } |4| = \sqrt{x^2 + y^2} = 2$$
$$|\nu| = \sqrt{2^2 + t^2} = 3$$
$$4 \cdot \nu = |4| \cdot |\nu| \cdot \cos\theta = 1 \Rightarrow \frac{t}{x} = \frac{z}{y} = \frac{|\nu|}{|4|} = \frac{3}{2}$$
$$4 \cdot \nu = x_t + y_z = 6$$
$$A = x + 1{,}5y \le (x^2 + y^2)(1^2 + 1{,}5^2) =$$
$$\sqrt{4 \cdot (1 + 2{,}25)} = \sqrt{13}$$

12. 2.

$$x^2 = (1 - x)^2 + \frac{1}{9}$$
$$x^2 = 1 - 2x + x^2 + \frac{1}{9}$$
$$2x = \frac{10}{9}$$
$$x = \frac{5}{9}$$
$$\Delta ACE \sim \Delta C'DT \Rightarrow$$
$$\frac{\frac{2}{3}}{1 - x} = \frac{z}{x} = \frac{y}{\frac{1}{3}}$$
$$z = \frac{2x}{3(1 - x)} = \frac{2 \cdot \frac{5}{9}}{3 \cdot \frac{4}{9}} = \frac{5}{6}$$
$$y = \frac{2}{9(1 - x)} = \frac{2}{9 \cdot \frac{4}{9}} = \frac{1}{2}$$
$$PAEC' = \frac{5}{6} + \frac{1}{2} + \frac{2}{3} = \frac{5 + 3 + 4}{6} = \frac{12}{6} = 2$$

MathTopia Press

Problem Set B

1. D.
2. 240.
3. $\frac{7}{4}$.
4. B.
5. 8.
6. 27.
7. 6.
8. 10.
9. $2\sqrt{10}$.
10. 120.
11. D.
12. $24\sqrt{3}$.

Practice-1

1. A.
2. B.
3. C.
4. E.
5. A.
6. E.
7. E.
8. ?.
9. D.
10. E.
11. D.
12. C.

Practice-2

1. C.
2. C.
3. A.
4. C.
5. D.
6. C.
7. D.
8. B.
9. D.
10. D.
11. C.
12. C.

Practice-3

1. E.
2. C.
3. B.
4. B.
5. D.
6. 405.
7. D.
8. D.
9. E.
10. 106.
11. C.
12. E.

BOOK & WEB RESOURCES

In the preparation of GeoTopia 0.5, a variety of textbooks, instructional materials, and online resources were consulted as references and sources of inspiration. These materials represent diverse approaches to geometry, including Euclidean foundations, problem solving, proofbased reasoning, and competition-style challenges. The following references are provided for readers who wish to further explore related perspectives and resources.

Selected References

- Alexander, D. C., & Koeberlein, G. M. Elementary Geometry for College Students.
- Bassarear, T. Mathematics for Elementary School Teachers, 5th ed. (2012).
- Kanbir, S., & Buber, Z. (2023). Mathematical Fluency: Mastering Essential Mathematical Concepts and Skills, Grades 6–8. WI, USA: MathTopia Press.
- Kanbir, S., & Tetikoglu, S. F. (2021). GeoTopia Volume 1: Learning Geometry Through Proofs and Challenging Problems. WI, USA: MathTopia Press.
- McDougal Littell / Houghton Mifflin. Geometry for Enjoyment and Challenge.
- Southall, E. Geometry Juniors.
- Waite, E., & Leff, L. S. Barron's Math 360: A Complete Study Guide to Geometry.

Online Mathematics and Research Resources

- K12 Math Contests: https://www.k12mathcontests.com/
- Brilliant: https://brilliant.org/
- Illustrative Mathematics: https://illustrativemathematics.org/
- New York State Regents Geometry Examinations: https://www.nysedregents.org/geometryre/
- Art of Problem Solving (AoPS): https://artofproblemsolving.com
- OpenAI ChatGPT (GPT–5): https://chat.openai.com

Online resources, including computational and language-based tools, were used for general exploration, drafting support, and idea generation. Final mathematical content, proofs, and problem solutions remain the responsibility of the author.

Made in United States
North Haven, CT
04 November 2026

21774382R00089